Paint Formulation

Principles and Practice

Paint Formulation

Principles and Practice

J. Boxall
and
J. A. von Fraunhofer

George Godwin Limited London

First published in Great Britain 1980 by
George Godwin Limited
The book publishing subsidiary of
The Builder Group
1–3 Pemberton Row, Red Lion Court,
Fleet Street, London EC4P 4HL

British Library Cataloguing in Publication Data

Boxall, John
Paint formulation.
1. Paint
I. Title II. Von Fraunhofer, Joseph Anthony
667'.6 TP936

ISBN 0-7114-5519-8

Printed in Great Britain by
The Pitman Press, Bath

Preface

The satisfactory formulation of a paint is a complex subject: the demands imposed on paint coatings are many and varied. While it is a relatively simple matter for the architect, designer or protection specialist to specify the properties he requires of a coating, the actual formulation of a paint to meet these requirements is far more difficult. A number of considerations must be borne in mind, such as the nature of the substrate, the proposed service environment and the method of application.

This book has been written in response to the need for a standard text covering the essentials of paint formulation. The approach adopted is to present the fundamental principles and then to apply them to actual cases, with examples of paint formulations for the uses and substrates discussed. These are all usable formulations and they provide the basis for a wide variety of coatings.

Practising paint technologists and students will find the book a convenient guide to the subject. Moreover, the use of chemical terms rather than trade or proprietary names, together with attention to underlying principles, allows it to serve both as a working manual and a reference handbook.

It has been necessary to summarise and simplify to some extent when dealing with types of paint component, types of coating for specific substrates and actual example formulations. This course has been followed deliberately, since a rigid presentation in which every item is minutely and precisely specified is likely to stifle initiative and give a misleading impression with regard to the wide variety available to the formulator.

With logic, and on the basis of the principles outlined in the book, a successful coating system can be formulated for any given substrate–service environment using a number of different approaches. However, the formulations presented here should not be taken as invariable standards, since new and modified materials are continuously becoming available: improvements are always possible.

J. Boxall
J. A. von Fraunhofer

January 1980

To Friendship

What we have to learn to do, we learn by doing: Aristotle

Contents

Preface v

1 Basic Paint Technology 1

1.1 Introduction 1

1.2 Polymers 2
1.2.1 Convertible coatings 2
1.2.2 Non-convertible coatings 10
1.2.3 Miscellaneous binders 12

1.3 Pigments 14
1.3.1 General considerations 15
1.3.2 Colouring pigments 15
1.3.3 Corrosion-inhibitory pigments 20

1.4 Extenders 23

1.5 Solvents 24
1.5.1 Hydrocarbon solvents 27
1.5.2 Alcohols and ethers 27
1.5.3 Esters and ketones 28

1.6 Plasticisers 28

1.7 Driers 30

1.8 Bioactive additives 30

1.9 Viscosity modifiers and anti-settling agents 32

1.10 Pigment-dispersing agents 33

1.11 Bibliography 34

2 Essential Concepts of Paint Formulation 35

2.1 Introduction 35

2.2 Pigment loading 36
2.2.1 Introduction 36
2.2.2 Pigment-to-binder ratio 36
2.2.3 Pigment volume concentration (P.V.C.) 37

2.3 The solvent phase 42
2.3.1 Introduction 42
2.3.2 Evaporation rate 43
2.3.3 Solvent power 44
2.3.4 Flash-point 45

2.4 Viscosity 45
2.4.1 Introduction 45
2.4.2 Application and storage considerations 45
2.4.3 Viscosity control 48
2.4.4 Viscosity measurement 49

2.5 Methods of formulation 49
2.5.1 Introduction 49
2.5.2 Statistical methods 51
2.5.3 Non-statistical methods 51

2.6 Pigment dispersion 53
2.6.1 Introduction 53
2.6.2 Mechanism of dispersion 54
2.6.3 Surface-active agents 54
2.6.4 Paint manufacture 56
2.6.5 Assessment of dispersion 58

2.7 Colour technology 58
2.7.1 Introduction 58
2.7.2 Colour matching 58
2.7.3 Mass colour and tint strength 59
2.7.4 Lightfastness 60
2.7.5 Stability 60

2.8 Bibliography 61

3 Mathematics in Paint Formulation 63

3.1 Paint formulae 63
3.2 Specific gravity 65

3.2.1 Cumulative specific gravities 65
3.2.2 Specific gravity of a polymer in solution 66

3.3 Drier additions 66

3.4 Percentage volume solids 67

3.5 Paint application rate and film thickness 69

3.6 Paint requirements 70

3.7 Pigment-to-binder ratio 70

3.8 Pigment volume concentration 71
3.8.1 Specific gravity of pigments 72
3.8.2 Oil absorption 73

3.9 Basic statistics 73
3.9.1 Introduction 73
3.9.2 Averages and means 75
3.9.3 Calculation of means 75
3.9.4 Variation 76
3.9.5 Normal curve and probabilities 78

3.10 Statistical tests 80
3.10.1 Introduction 80
3.10.2 Standard error of the mean 81
3.10.3 Student's *t* test 83
3.10.4 Analysis of variance 85
3.10.5 Other statistical tests 87

3.11 Bibliography 89

4 Formulation of Coatings for Steelwork 91

4.1 Introduction 91

4.2 Paint systems for steel 92

4.3 Industrial coatings 94
4.3.1 Industrial coatings for mild conditions 94
4.3.2 Industrial coatings for moderate conditions 97
4.3.3 Industrial coatings for severe conditions 103

4.4 Site-applied coatings 107
4.4.1 Site-applied coatings for mild conditions 107
4.4.2 Site-applied coatings for moderate conditions 112
4.4.3 Site-applied coatings for severe conditions 115

4.5 Bibliography 124

5 Formulation of Coatings for Timber 125

5.1 Introduction 125

5.2 Paint systems for timber 127

5.3 Coatings for exterior timbers 129
5.3.1 Types of paint system 129
5.3.2 Primers 129
5.3.3 Finishes 132

5.4 Coatings for interior timbers 136
5.4.1 Varnishes and lacquers 136

5.5 Bibliography 144

6 Formulation of Coatings for Masonry 145

6.1 Introduction 145

6.2 Paint systems for cementitious surfaces 146

6.3 Coatings for interior substrates 147

6.4 Coatings for exterior substrates 147
6.4.1 Exterior substrates: fresh, wet 147
6.4.2 Exterior substrates: fresh, damp 148
6.4.3 Exterior substrates: fresh, dry 152
6.4.4 Exterior substrates: weathered 158

6.5 Bibliography 161

7 Miscellaneous Substrates 163

7.1 Introduction 163

7.2 Formulation of coatings for plaster 163
7.2.1 Introduction 163
7.2.2 Plaster as a paint substrate 164
7.2.3 Paints for plaster 165

7.3 Formulation of coatings for non-ferrous metals 168
7.3.1 Introduction 168
7.3.2 Paints for aluminium 169
7.3.3 Paints for zinc 170

7.4 Formulation of coatings for plastics 171
7.4.1 Introduction 171
7.4.2 Paints for plastics 171
7.5 Bibliography 174

Appendix 1: Conversion Tables 175

Appendix 2: British Standards 183

Index 187

Publisher's note

CHAPTER 1

Basic Paint Technology

1.1 Introduction

Paints are surface coatings that are designed to fulfil a protective and /or decorative function for the substrate. While inorganic paints are known, organic coatings predominate in paint technology and this is reflected in the coverage of this book.

The formulation of a paint in respect of its major components, the resinous binder and the pigments, is largely dictated by the intended application, i.e. by the combined requirements of the substrate and the service environment. Thus paint systems for metals, timber, masonry, etc., may differ markedly from each other, and coatings designed primarily to protect can be formulated on different principles from those that apply to coatings whose function is largely cosmetic. The factors important in formulating paints are considered in later chapters.

The major components of a paint formulation are the binder (including any added plasticiser), the pigments (as well as any extender material), a solvent or blend of solvents, and various other substances which may be added at relatively low levels but have a marked effect on the paint. Amongst these auxiliary additives can be included thickeners, driers, anti-skin agents, surface-active agents, biocides, fungicides and numerous others.

The interrelation of all these various components and additives is discussed in detail in this book but some consideration of their characteristics may be useful as an introduction to the principles of paint formulation. The basic physics and chemistry of the resins, pigments, solvents, etc., used in paint technology are treated in some detail in a number of specialised monographs (see bibliography at the end of the chapter).

In this chapter the properties and functions of several of the more important raw materials used in paint formulation are discussed. Attention is devoted to basic theoretical considerations, examples being given of certain of the components used to produce the desired paint characteristics.

1.2 Polymers

The polymeric or resin binder of a paint is the film-forming component of the formulation; without it, continuous coatings would not be possible. However, a large variety of polymers (resins) and polymer-precursors (materials that will undergo reaction to form a resin) are used in paints. Furthermore, the chemical nature of the material dictates its mode of use and, indeed, the overall formulation of paints based on the resin (or resin combination).

Paint binders can be subdivided into two broad categories: convertible and non-convertible types. The former are materials that are used in an unpolymerised or partially-polymerised state and which, following application to the substrate, undergo reaction (polymerisation) to form a solid film. Non-convertible paints are based on polymerised binders dispersed or dissolved in a medium which evaporates after the coating has been applied, to leave a coherent film on the substrate surface.

Some examples of convertible binders to be considered here include oils, oleoresinous varnishes, and alkyd, amino, epoxy, phenolic, polyurethane and silicone resins. Non-convertible binders include cellulose, chlorinated rubber, acrylic and vinyl resins.

1.2.1 CONVERTIBLE COATINGS

Oils and oleoresinous varnishes

The use of oils in paint formulation was widespread throughout historical times and until the mid-20th century. More recently, however, with the introduction of more sophisticated polymers, their use has declined markedly so that, with the exception of a limited use of refined linseed oil and linseed stand oil in certain types of primer for steel and timber, oils as such are rarely encountered. They are extensively used, however, as modifying components in the preparation of oleoresinous varnishes and especially in the oil-modified alkyd resins, the latter forming the most widely used class of binder in modern paint technology.

Vegetable oils are generally classified as either drying or non-drying oils on the basis of whether they will react with air at ambient temperatures to form a rubbery, poorly solvent-soluble film. Typical drying oils are linseed and tung, both of which have pronounced drying properties. Other oils, such as soya bean oil, will also dry but less readily, and these are known as semi-drying oils. In contrast, oils such as castor and olive do not dry and form films except over very long periods and, consequently, are known as non-drying oils. It is possible, however, to convert a non-drying oil such as castor oil into a drying oil, e.g. dehydrated castor oil (D.C.O.). In this particular case conversion is achieved by heating at 280°C for several hours, when there is a decomposition of the constituent acids of the oil.

The drying process is a complex oxidation reaction involving the centres of unsaturation in the carbon chains of the fatty acid triglycerides of the drying oil, the reaction being known as oxidative polymerisation or auto-oxidation. Although oxidation of a drying oil, and hence film-formation, will occur naturally, the rate of reaction is slow and is usually accelerated by the addition of driers, for example cobalt or lead naphthenates.

As previously stated, linseed oil is still used in its unmodified state (that is, unmodified with other resinous materials), although certain treatments are necessary to render it suitable for use. The crude linseed oil extracted from the seeds of the flax plant is generally dull and cloudy and needs refining to remove the causative compounds, i.e. the natural impurities and particles of meal. The method of refining depends largely on the purpose for which the oil is required; treatments involving alkalis or acids can be used, as well as bleaching processes to lighten the colour of the derived films. For example, alkali-refined linseed oil can be used for the production of varnishes where the low acid values obtained are beneficial, whereas acid-refined oils can be used in paints, as the higher acid value facilitates pigment dispersion processes.

Alkali refining is the more commonly used method. In this process the oil is treated with sodium hydroxide solution in a calculated quantity to neutralise the acid value of the oil (the acid value is the numerical measure of the free acids in a polymer and is quoted as the number of milligrammes of potassium hydroxide required to neutralise the free acids in 1 gramme of the substance). The resultant soap, formed of acids and other impurities as well as a small proportion of the oil, is removed by centrifuging. The refined oil is then washed with water to remove all traces of the alkaline material.

Refined linseed oils are generally of low viscosity, however, and the use of heat treatment produces thicker materials known as linseed stand oils. Typically, the process involves heating the previously refined oil to about 270°–300°C for a time depending on the ultimate viscosity required. The oil treated by this method thickens mainly by polymerisation, and the film properties of flexibility and durability, as well as flow of the liquid varnish, are improved compared with refined linseed oil. Drying times, however, tend to be longer.

In varnishes, many types of oil are utilised, although in general these tend to be of the drying oil variety such as linseed, tung or dehydrated castor oil. The method of manufacture varies depending on the precise nature of the components used; typically, the oil is heated and a resin component is slowly added. Heating is continued until solution is effected; the viscosity is then reduced to an appropriate level with a suitable solvent, usually white spirit.

The various resins that can be used differ widely in their composition and characteristics and this clearly has a major influence on the end use of the derived varnish products. The resins used can be either natural (fossil) resins such as copal or kauri, or (and now to a greater extent) synthetic materials

Table 1.1 Oleoresinous Varnishes

Oil length	Oil : resin ratio	Characteristics	Uses
Short	~0·5–1·5 : 1·0	Fast drying, but hard brittle films	Fillers Interior varnishes
Medium	1·5–3·0 : 1·0	Slower drying, less hard films	Paints and varnishes
Long	3·0 – ~ 5·0 : 1·0	Slow drying, more flexible and durable films	Exterior paints and varnishes

such as rosin ester gum, coumarone, rosin-modified phenolic and pure phenolic resins. Perhaps the most important types are the phenolic-based materials (see page 8), used in conjunction with oils such as tung or linseed.

In contrast to the properties of films derived from drying oils, varnishes generally tend to dry more rapidly (although driers are still required), and they possess superior hardness, gloss and flow properties. With certain types, such as shorter-oil-length materials and rosin-containing products, the durability is reduced. The ultimate properties of a varnish are determined not only by its composition in respect of ingredients, but also by its oil length, that is the ratio of oil to resin (Table 1.1).

Alkyd resins

Alkyd resins constitute the most widely used paint binders; unlike oleoresinous materials, they are synthetic polyesters. They are manufactured by the interaction of a polycarboxylic or fatty acid or its anhydride (e.g. phthalic anhydride), a polyhydric alcohol (e.g. glycerol) and a vegetable oil or its fatty acid.

Two manufacturing methods are used in the preparation of alkyd resins: the alcoholysis (or monoglyceride) process, by which oil-modified alkyds are produced, and the fatty acid process which, as its name implies, is used to prepare fatty-acid-modified alkyd resins. These latter resins tend to have a paler colour than their oil-modified counterparts, due to the greater purity of the fatty acids.

In the fatty acid process, all three components, the fatty acid, the polyhydric alcohol and the polycarboxylic acid, are heated together at temperatures in the region of 240°C until the esterification reaction is complete. In the alcoholysis process, the polyhydric alcohol and modifying oil are first reacted together and then the polycarboxylic acid is added.

Both processes yield oil-modified alkyds and these are classified on the basis of the amount (i.e. oil length) and type of oil used (Table 1.2). The medium- and short-oil (length) alkyds are often based on semi- and

Table 1.2 Oil-modified Alkyds

Alkyd	Oil length (%)	Characteristics	Uses
Short	20 – 45	Non oxidative Soluble in aromatic hydrocarbon solvents Hard brittle films	As modifying resins in stoving systems for interior use
Medium	45 – 60	Oxidative (air-drying) or stoving cured Soluble in aliphatic–aromatic hydrocarbon solvent blends More flexible films	As above for interior–exterior applications Also in quick air-drying systems
Long	60 – 80	Oxidative Soluble in aliphatic hydrocarbon solvent blends More flexible films	Exterior air-drying systems

non-drying oils, while both semi-drying and drying oils (or their constituent acids) are used for the long-oil alkyds. Typical examples of drying oils used in alkyds include linseed and dehydrated castor oil, while soya bean oil is representative of semi-drying oils, and castor and coconut are examples of non-drying oils.

Drying and semi-drying oil alkyds cure (form films) by auto-oxidation, whereas a stoving regimen is required for non-drying oil alkyds, particularly the short-oil resins; alternatively they may be used as plasticisers for other synthetic resin binders (see page 29).

As already stated, the underlying concept in the manufacture of oil-modified alkyd resins is that the free hydroxyl groups in an alkyd resin such as glyceryl phthalate can be reacted (esterified) with the fatty acids present in a vegetable oil. Alkyd resins are therefore essentially polyesters and as such are susceptible to degradation by strong acids and alkalis. This limits their use to situations where such exposure will not be experienced. However, oil- and fatty-acid-modified alkyds possess, with respect to general durability, ease of application and tolerance to adverse application conditions, attributes which render them especially suited to the formulation of general-purpose paint systems for the construction industry.

Modified alkyd resins are also used, prepared by means of either chemical modification or blending (see page 115); the preparation of styrene-modified alkyds illustrates the general technique of chemical modification.

Styrenation of alkyds, i.e. interreaction between the alkyd resin and the styrene in the presence of a peroxide catalyst, is a widely adopted procedure. This yields a copolymer in which the double bond in the ethylene side-chain of the styrene has reacted (or copolymerised) with the unsaturated centres in the fatty acids in the drying oil component of the alkyd resin. Styrenated alkyds have faster curing rates and greater chemical resistance than conventional alkyd resins. The characteristics of the styrenated alkyd vary with proportion of styrene used as well as with the nature of the alkyd resin precursor, the oil length and the oil type. Consequently, a wide variety of styrenated alkyd resins is possible through selection of the reaction conditions and the raw materials used. Other vinyl compounds such as vinyltoluene may also be used to modify alkyd resins.

Double bonds in the fatty acid components of alkyd resins can also be copolymerised with other resin precursors. Thus interaction with isocyanates will yield urethane alkyds, and various other copolymers are known and used.

Amino resins

Condensation polymerisation of an aldehyde with an amine or amide yields an amino resin, the most important being the urea-formaldehyde and melamine-formaldehyde polymers. These polymers are insoluble in the common solvents but they are rendered soluble by butylation. Butanol is added to the initial reaction mixture and the butylated amino resins are solvent-soluble. The resins will form films, either through the addition of small amounts of a strong acid or by heating at 100°–150°C. The resulting films tend to be very brittle, therefore amino resins are normally plasticised with alkyd resins.

A non-drying-oil alkyd–amino resin blend yields a stoving cured resin; the addition level of the non-drying-oil alkyd can be up to 50 per cent. In contrast, with the drying-oil alkyds, up to 10 per cent resin is used as a hardening additive. The amino resin accelerates the hardening of the film, although a stoving regimen is used for curing. The resultant films are hard with good gloss, flexibility and durability as well as exhibiting good adhesion to metals and water resistance. The urea resins have greater solvent solubility than the melamine resins, although the latter have a more marked hardening effect and so can be stoved at somewhat lower temperatures, while the films have superior properties.

Amino resins are important constituents of stoving finishes but are little used on their own except in the general field of polymer technology, typically for industrial moulding applications.

Epoxy resins

Epoxy resins are produced by condensation polymerisation between epichlorhydrin and diphenylol-propane (bis-phenol A), usually in the presence of sodium hydroxide and under reflux. The reaction conditions and the proportions of the two constituents determine the properties of the final product, a linear polymer of low solubility in the common (i.e. non-polar) solvents but soluble in highly polar solvents such as ketones.

The epoxy resins have glycidyl ether or oxirane terminations and these cyclic epoxides can open so that the resin can undergo cross-linkage. This converts them from linear thermoplastic materials to three-dimensional thermosetting resins. This reaction is slow, even with heat, but a number of reagents, particularly tertiary amines, accelerate ring opening and promote cross-linkage without heating. Tertiary amine salts are often used in preference to the amines themselves since they permit greater additions of the 'catalyst' to be made without affecting the pot life. Polyfunctional amines, e.g. *m*-phenylenediamine, are also used as curing agents. In general, aliphatic amines effect cure at room temperature and give faster reaction rates, while aromatic amines give resins with higher T_g values. Amines, however, can be toxic or cause adverse skin reactions so that there is now a trend to use amine adducts or polyamides as curing agents.

Since epoxy resins will undergo cross-linkage in the presence of amines, polyamines and amine adducts at room temperature, they are referred to as cold-curing materials. The two components need to be packaged separately and mixed just prior to use.

Epoxy resins may be blended with other polymers such as acrylic, amino and phenolic resins and, after cure, the cross-linked material possesses the characteristics of both precursors. However, the epoxy–phenolic and epoxy–amino resin blends require heat (about 200°C) to effect cross-linkage; with the amino blends, acidic catalysts can reduce the cure temperature to about 150°C. Both baking finishes are characterised by good resistance to chemicals and heat as well as colour stability. Epoxy–acrylic resin blends are cold-curing and have good aesthetic properties but their chemical resistance and thermal stability are inferior to those of the other (heat-cured) blends.

Coal tar is often blended with epoxy resins and epoxy copolymers to upgrade their resistance to water; coal tar–epoxy coatings are used in many marine applications.

Reaction of epoxy resins with the fatty acids in vegetable oils yields the epoxy esters, and the characteristics of films formed from these binders are determined by the fatty acid and the oil length. Epoxy esters are similar to alkyd resins but, since the ester linkages are in sidechains and not in the backbone of the resin as in alkyds, the epoxy esters have greater chemical resistance. Drying, semi-drying and non-drying oils can be used to prepare epoxy esters and, as with the oleoresinous varnishes and alkyds, classification is on the basis of oil-length. Long-oil epoxy esters based on drying and semi-drying oils are air-drying, while the medium- or short-oil epoxy esters based on drying or semi-drying oils require baking. Short semi-drying and non-drying oil epoxy esters also require baking and are usually blended with amino resins, the latter conferring greater hardness and chemical resistance upon the film.

Phenolic resins

When a phenol (i.e. an aromatic compound having a hydroxyl group attached to the benzene nucleus) reacts with an aldehyde such as formaldehyde in the presence of a catalyst, low-molecular-weight resins are formed. If the phenol is in excess, novolak resins are obtained; resol resins are produced if the aldehyde is in excess. When heated, both resins undergo condensation polymerisation and cross-linkage if a cross-linking agent such as hexamethylenetetramine is present.

Resol and novolak resins are little used in surface coatings because they are poorly soluble in oil. If, however, a substituted phenol such as *p*-tert-butylphenol is used to prepare the resin, the resultant resin is oil-soluble. Varnishes prepared from these '100 per cent phenolic' resins have greater durability than the rosin-modified phenolic resins. The latter are prepared by heating resol resins with excess rosin, this yielding a mixture of complex acids which can be esterified with a polyol such as glycerol. These esters can be heated with drying oils to yield binders. Long-oil combinations are air-drying, whereas the short-oil combinations are cured by stoving.

Coatings based on rosin-modified phenolic resin–drying oil combinations have good chemical resistance but have a tendency toward brittleness and poor exterior durability.

Polyurethane resins

Polyurethane resins are polymers formed by the reaction of isocyanates with hydroxy compounds. The resins of interest to paint technology are derived from di-isocyanates, typically toluene di-isocyanate, diphenylmethane di-isocyanate and hexamethylene di-isocyanate. If the di-isocyanate is reacted with a diol (aliphatic glycol), a linear polymer is formed; reaction with water and polyols (including some vegetable oils, polyesters and polyethers) yields cross-linked polymers.

Polyurethane resin binders fall into two broad categories, single-pack and two-pack. Three types of single-pack system are known: air-curing, moisture-curing and heat-curing or stoving. The two-pack systems are cold-curing, that is, they polymerise in the presence of a catalyst.

Moisture-curing polyurethane systems consist of prepolymers having isocyanate terminations to the main chain, these being formed by reaction of an isocyanate with a polyhydric alcohol. The isocyanate terminal group will react with atmospheric moisture and so initiate a series of reactions which result in polymerisation and film formation. The final films are hard and flexible with high chemical and abrasion resistance.

The heat-curing systems also consist of prepolymers formed by reaction of an isocyanate and a polyol, but differ in that the isocyanate terminations are blocked by reacting them with phenol. The system remains blocked or non-reactive until heated, when the phenol is lost and the isocyanate termination can react with hydroxyl groups present in the formulation to effect cross-linkage. The resultant films have excellent physical and chemical properties and the overall system has high storage stability.

The air-curing single-pack urethane systems are known as urethane-alkyds (or uralkyds). These binders are akin to the conventional alkyds and are prepared in a similar way except that an isocyanate is used instead of the carboxylic acid in alkyd manufacture. The type and length of oil used to prepare the uralkyd determine the film properties and the films are characterised by good resistance to water, mild corrosives and abrasion.

The two-pack urethane systems are based either on an isocyanate–polyol prepolymer or an isocyanate adduct–polyol combination. The isocyanate–polyol system comprises an isocyanate-terminated prepolymer in one pack and a catalyst, such as a tertiary amine, in the other pack. When mixed, the components react in the cold but often stoving is employed to accelerate the cure rate, resulting in fairly densely cross-linked resins. The films have excellent properties and the pot life of the mixed components is quite long. The isocyanate adduct system consists of a non-volatile isocyanate, i.e. the reaction product of an isocyanate with a polyol, a polymerised isocyanate (a polycyanurate) or a low-volatility isocyanate, as one component. The second component of the system is a polyester or polyether containing free hydroxyl groups. When mixed, the two components react and cross-linkage occurs. Because the isocyanate component tends to react with atmospheric water vapour or adsorbed moisture on the substrate (and so leads to softer films), a slight excess of this second component is incorporated, although too high a level leads to film brittleness. The resultant films again have high resistance to chemicals and abrasion, excellent physical properties and good durability.

Silicone resins

Silicone resins differ from the majority of other paint polymers in that they have an inorganic backbone, i.e. alternating atoms of silicon and oxygen. Organic groups are attached to the silicon atoms and in fact three types of silicone polymer exist, namely rubbers, fluids and resins, of which only resins are of importance in paint technology. It should be noted that these silicon–oxygen polymers have the generic name of siloxanes, but are commonly known as silicones.

Silicone resins are produced by hydrolysis of di- and trichlorosilanes, the resultant silanols undergoing spontaneous condensation. Methyl and phenyl chlorosilanes are used to form the silicone resins which are a mixture of linear and cross-linked polymers. Upon stoving, the resins undergo further condensation and form more highly cross-linked structures.

The proportions of methyl and phenyl radicals affect the resin properties: the methyl yields hard but weak films, while the phenyl groups give stronger but brittle films. The dichlorosilanes are included to moderate the degree of cross-linkage.

Pure silicone resins are used only for stoving finishes, when they yield films with excellent resistance to aqueous media and chemicals. Silicone resins can be blended or copolymerised with other resins such as alkyds and polyurethanes to yield films of enhanced durability, water-repellency and temperature-resistance.

1.2.2 NON-CONVERTIBLE COATINGS

Cellulosic polymers

Cellulose is the main constituent of the cell walls of all plants and is the most widely occurring of all natural polymers. The primary sources of industrial cellulose are wood and cotton fibre. The material itself is insoluble but many of its derivatives are soluble and utilised, for instance, in cellulosic lacquers.

Cellulose nitrate (or, commonly, nitrocellulose) is produced by the action of nitric and sulphuric acids on cellulose followed by hydrolysis with water at elevated temperatures and pressures. The reaction product is dehydrated with ethanol under pressure to yield a material containing 11–12 per cent nitrogen, 5 per cent water and 30 per cent alcohol.

Cellulose nitrate is soluble in a wide range of solvents and film formation is by solvent evaporation, although plasticisers are required to provide film flexibility and adhesion. The films are resistant to water and dilute acids but are decomposed by alkalis and concentrated acids. Until recent years, cellulose nitrate lacquers were of major importance for the fast-drying finishes used on motor vehicle bodies.

Chlorinated rubber

Natural rubber is not used in paint technology but its derivatives, notably chlorinated rubbers (and, to a smaller extent, isomerised rubber), are important.

Chlorinated rubber is prepared by chlorination, at about 80°–100°C, of masticated crepe rubber dissolved in chloroform or carbon tetrachloride, the reaction product containing some 60–65 per cent combined chlorine.

The chlorinated rubbers may be used as paint binders, when they form films by solvent evaporation from solutions in aromatic hydrocarbons; fairly high plasticiser levels are required to provide film flexibility. The materials are also blended with many other resins to confer water resistance and chemical resistance upon the binder.

Chlorinated rubber binders are primarily used for coatings required to provide high resistance to chemicals and/or corrosion.

Vinyl resins

Vinyl resins are polymers derived from monosubstituted ethenes, i.e. ethylene with one hydrogen replaced by the hydroxyl, chloride, acetate or other groups. Of these, poly(vinyl chloride), p.v.c., and poly(vinyl acetate), p.v.a., have the greatest importance in paint technology, although other vinyl resins, particularly various copolymers, are also used as binders.

Four preparative routes are used to polymerise vinyl compounds: bulk, suspension, emulsion and solution polymerisation. In bulk polymerisation, the vinyl compound, together with a catalyst such as benzoyl peroxide, is heated for 10–12 hours at 60°C. This is performed in a rotating sealed drum containing stainless steel balls and the particle size of the polymer is determined by the ball diameter and drum rotation speed. In suspension polymerisation, the vinyl compound is suspended in water together with a suitable initiator as the polymer precursor, the suspension being heated at 50°C under pressure for several hours to effect polymerisation. The resin product is obtained as a slurry which is separated by filtration and drying, the polymer particle size and shape being determined by the suspension agent used. Emulsion polymerisation also utilises pressure treatment at 50°C but differs in that the precursor is an emulsion of the vinyl compound in water and has a faster reaction rate. The vinyl resin is obtained as a latex which, after spray-drying, yields small spherical particles. In solution polymerisation, the vinyl compound and catalyst are dissolved in a solvent as the precursor, and polymerisation proceeds in solution. The resultant vinyl resin is not separated from its reaction medium since the same solvent is used as its carrier for surface coatings.

The most important vinyl resin is poly(vinyl chloride) which is a colourless, rigid, thermoplastic polymer having excellent resistance to acids, alkalis, chemicals, aqueous media and oxidising agents. However, p.v.c. has a linear isotactic structure (i.e. all the chlorine atoms are on the same side of the carbon chain) and thus has a tendency to crystallise. This results in a limited solubility in the common solvents, but p.v.c. is soluble in chlorinated and oxygen-containing solvents (ketones, ethers and aromatic nitro-compounds). The poor solubility of p.v.c. limits its use in paints although, when dispersed in hydrocarbon or plasticiser solvating media (i.e. organosols or plastisols, respectively), it finds applications in surface coatings. When the dispersion is sprayed onto a substrate, discrete particles are deposited; these coalesce to form a continuous coating upon stoving. More commonly, p.v.c. is used in paints as a copolymer with poly(vinyl acetate), with ratios of vinyl chloride to vinyl acetate of 80:20 to 90:10.

Poly(vinyl chloride–vinyl acetate) copolymers can be prepared by esentially the same methods as the single resins although solution polymerisation is perhaps used most widely. The p.v.c.–p.v.a. copolymer is also a linear thermoplastic resin having a greater solvent solubility but slightly inferior mechanical properties and chemical resistance, although the latter is still good. Coatings based on the copolymer air-dry by solvent loss; stoving regimens are also used.

The adhesion of p.v.c.–p.v.a. films to metallic substrates may be improved by the incorporation of polar groups. This is achieved by low additions (about 1 per cent) of carboxylic acids, such as maleic acid or its anhydride, to yield terpolymers. Copolymerisation of vinyl chloride with maleic acid also yields resins of improved adhesion with little detriment to the properties of the resin.

Polymerised vinyl alcohol as such is not used in paint binders. However, when vinyl alcohol is hydrolysed and suspended in alcohol with the admixture of butyraldehyde and sulphuric acid as a catalyst, the mix then being heated for several hours, the copolymer poly(vinyl butyral) is obtained. This resin is used in the preparation of the so-called 'self-etch' primers and deposits tough, flexible films having good adhesion to metals. Resistance to aliphatic hydrocarbons and oils is good but the films are susceptible to the action of acids and strong bases, and have a tendency to soften in water.

Vinylidene chloride may be polymerised to a resin (p.v.d.c.) which, strictly, is an acrylic rather than a vinyl polymer — it is included here for convenience. Vinylidene chloride forms dense, crystalline resins but copolymerisation with vinyl chloride yields resins that deposit more flexible films. A typical v.d.c.–v.c. ratio is 85:15 and the copolymer is used as a binder where high resistance to aqueous media is required, as in swimming pools.

Acrylic resins

Acrylic resins are poly(vinylidene) compounds (see above); the best known is poly(methylmethacrylate) or p.m.m.a. The resins are derived from the esters of acrylic and methacrylic acid which are polymerised by a vinyl-type process (see page 11).

Acrylic resins are hard and rigid thermoplastic polymers that form somewhat inflexible films but which have good optical properties and excellent chemical resistance. They are attacked, however, by concentrated acids and oxidising agents. Film flexibility can be enhanced by copolymerisation: in the case of p.m.m.a., by means of ethyl or other acrylates. Cross-linkage can be effected by glycol dimethacrylate and cross-linked resins have improved chemical and physical properties while also being thermosetting and having lower solvent solubilities.

Copolymerisation of methyl (and other) methacrylates with monomers such as styrene, acrylonitrile and vinyltoluene also yields thermosetting resins.

Acrylic coatings are commonly based on dispersions of the resin in solvents or water. Self-cross-linking binders are also used, often in conjunction with other resin systems.

1.2.3 MISCELLANEOUS BINDERS

Two other binders have importance in paint technology although they are not strictly classified as polymers, namely bitumens (USA: asphalts) and inorganic silicates.

Bitumen

Bitumens, used in bituminous paints, are naturally occurring or distilled crude or petroleum oil residues. The grade of bitumen, i.e. its colour and properties, depends on both the source and distillation conditions. Naturally occurring bitumens (often known as asphalts) have high softening-points and the distilled bitumens are air-oxidised to raise their softening temperature. When dissolved in hydrocarbon solvents, bitumens form surface coatings by solvent evaporation. The coatings have good resistance to aqueous media, to chemicals, to moderate concentrations of non-oxidising acids and to bases at ordinary temperatures. Blending of bitumen with media that contain drying-oil results in paints with compromise properties, that is, rapid drying with good protective characteristics for the (metal) substrate under moderate-to-severe exposure conditions. The so-called coal tar–epoxy resin coatings are obtained by blending of bitumens with epoxy resins, which improves the epoxy resin binder's properties of impermeability to water and resistance to chemicals.

Inorganic silicates

Inorganic silicates are prepared by fusion of sand and silica with sodium and/or potassium carbonate, the fused mass then being digested in water. Zinc dust with zinc or calcium oxide is added and, when the solution is applied to a substrate, a silicic acid gel is formed by evaporation of water and interaction with atmospheric carbon dioxide. The silicic acid reacts with the zinc or calcium oxide (and, to a degree, with the zinc dust) to form complex inorganic silicates, additions such as potassium phosphates and lithium silicate being made to accelerate the curing rate. Such zinc dust–inorganic silicate coatings have exceptional durability, temperature resistance and corrosion resistance, but they may be attacked by acids and strong bases. They dry poorly under wet conditions but their aqueous resistance, especially to sea water, is exceptional and these coatings are used as protective primers for marine structures. The system is a two-pack formulation.

Organic–inorganic silicates consist of zinc dust mixed with ethyl orthosilicate which is combined with the inorganic silicate system. Both single- and two-pack formulations have been developed. The coating properties are superior to those of the inorganic silicates and they have a longer pot life.

A wide variety of binders may be utilised in paint technology, the selection of a particular material being based primarily on the anticipated service environment and substrate, together with such factors as the pigments to be

Table 1.3 Specific Gravity of Polymers

Polymer	Specific gravity (approximate)
Oils	0·95
Oil-modified alkyds	1·0
Aminos	1·25
Epoxies	1·2
Phenolics	1·2
Polyurethanes	1·2
Silicones	1·1
Cellulose nitrate	1·7
Chlorinated rubber	1·7
Vinyls	1·2
Acrylics	1·2
Bitumens	1·0
Inorganic silicates	2·7

used in the coating formulation. An important requirement for the calculation of the pigment volume concentration of coating formulations is the specific gravity of the binder solids. Some typical values are given in Table 1.3, but there can be a relatively wide range for some types of resins.

Many of these points are considered further in later chapters. Paint additives, particularly pigments, extenders and solvents, are discussed below.

1.3 Pigments

Pigments are particulate solids that are dispersed in paints to confer certain characteristics upon them. These characteristics include colour, opacity, durability, mechanical strength and corrosion protection for metallic substrates. In order to achieve the desired aim, pigments have to possess certain properties and these, together with various other characteristics, will be considered here.

Pigments fall into two principal groups, namely pigments and extenders:

(a) *Pigments* are primarily required to confer colour and opacity upon the binder but may also have to fulfil other functions such as providing corrosion-inhibiting characteristics. They may be subdivided into two main chemical classes, inorganic and organic dyestuffs, which differ markedly in their characteristics and uses. Both types are widely used in surface coatings, although organic dyestuffs tend to find their major application in decorative (as opposed to protective) formulations.

(b) *Extenders* are materials that are insoluble in the binder and have virtually no colouring or opacifying function. They are incorporated in the formulation to modify the properties of the paint.

1.3.1 GENERAL CONSIDERATIONS

Most pigments (here taken to cover both 'pigments' and 'extenders') are crystalline and the crystal form often affects the characteristics of the pigment. The size and shape of the pigment particles are an important consideration since they affect the agglomeration or packing within the paint binder or matrix. Particle size also affects the final film gloss, settling of the pigment during storage and wetting by the binder. Other factors such as colour, tint strength, colourfastness and opacity are inherent characteristics of the pigment. Density is another important factor, affecting not only settling but also the volume of pigment for a given weight addition.

Pigments have to be dispersed in the binder and, in order to achieve the desired effect, they must remain in suspension or be easily dispersed again should settling occur. Since paints are manufactured by mechanically incorporating the pigments into the binder, typically by ball milling or a similar technique, not only the physical characteristics mentioned above are important, but also the texture and the mechanical strength of the pigment particles.

Thus a large number of factors must be taken into consideration when selecting pigments for paints: readers are referred to the specialised monographs listed in the bibliography to this chapter for a detailed discussion of pigments and pigmentation. For the purpose of this book, a somewhat simplistic approach has been adopted in the example formulations given. In most cases white paints have been formulated, but the principles of incorporating most types of pigmentation are covered where appropriate in the various chapters. It is believed that this will provide the information necessary for formulating a wide variety of coatings. Tables of some pigment constants are given in Chapter 2.

1.3.2 COLOURING PIGMENTS

A very wide variety of colouring agents is available for paints so that a considerable range of colours and shades is possible. Both inorganic and organic agents fulfil this function, the organic pigments being principally colouring additives while the inorganic materials also often have a second function which may sometimes be more important than their colouring effect.

White inorganic pigments

Titanium dioxide, TiO_2: a widely used, non-toxic, stable synthetic pigment used for both protective and decorative paints. Two crystallographic forms exist, rutile and anatase, the former being more compact and thus having a higher specific gravity, greater refractive index and enhanced stability and durability.

The high refractive index of titanium dioxide (2·7 for rutile and 2·5 for anatase) enables it to provide a higher degree of opacity in paint films than any other white pigment. Rutile titanium dioxide is photochemically inert and the pigment protects paint films from degradation by scattering absorbed light. In contrast, the anatase form is photochemically active and paints formulated with this pigment are subject to heavy chalking. Accordingly, the anatase grade is mainly used in interior situations where its cleaner white colour compared with the rutile form is also beneficial.

Zinc oxide, ZnO: a synthetic basic pigment which will inhibit mould growth if present at high levels (about 30 per cent by weight of the film). The basic nature of zinc oxide leads to interreaction with paints of high acid value with consequent formation of zinc soaps. These can mechanically reinforce the film but may cause embrittlement on exterior exposure.

Antimony oxide, Sb_2O_3: a synthetic, inert pigment widely used, in conjunction with chlorine-containing binders, in the formulation of fire-retardant paints. Upon exposure to flames, antimony chloride vapour is formed and blankets the fire. The pigment has good opacifying characteristics.

White lead, $2PbCO_3.Pb(OH)_2$: a synthetic basic pigment used for centuries but now in limited use because of its toxicity. Being basic, white lead will react with acidic paints to form lead soaps which will enhance the film elasticity, a useful attribute for timber primers. However, the pigment has a high chalking rate as well as a pronounced tendency to darken in polluted atmospheres through conversion to lead sulphide.

The other major white pigment, basic lead sulphate, is principally used in corrosion-inhibitory primers (see page 21).

Yellow inorganic pigments

Lead chromates, $PbCrO_4$ (lead chromes): a wide range of synthetic pigments that range from pale yellow to deep orange (and scarlet) in colour, but which are toxic. Lead chromes have high tint strengths, good lightfastness and opacity, but they darken in polluted atmospheres (containing hydrogen sulphide) and bleach in the presence of sulphur dioxide. Furthermore, lead chromes interact with alkaline substrates, resulting in colour loss. Nevertheless, the lead chromes are useful colouring pigments and are used in undercoats and finishing systems for both decorative and industrial coatings. The colour of the pigment is determined by its composition which in turn is controlled by the process conditions during manufacture.

Zinc chromates, $ZnCrO_4$: three forms of zinc chromate are used for pigmentation of paints. The colouring (pigmentary) grades have good lightfastness and are colour-stable towards both alkalis and sulphur dioxide but they are of low opacity. The zinc chromates are slightly basic and will react with acidic paints, leading to thickening of the liquid paint during storage. The form of zinc chromate used for corrosion-protection (see page 21) must be free from residual chloride, an impurity present in the purely colouring grade. A third form of zinc chromate, the tetroxychromate, is used in metal pretreatment primers (pages 21 and 99).

Yellow iron oxides, $Fe_2O_3.H_2O$: naturally occurring iron oxides, ranging in colour from dull yellow to dark yellow-brown, but more restricted in the range of colour shades than the synthetically prepared oxides. The latter also have cleaner and brighter colours combined with a greater colour strength by virtue of their greater purity.

Both natural and synthetic oxides are resistant to alkalis and organic acids although they are attacked by mineral acids, undergoing discoloration. Shade alteration can occur at elevated temperatures. Iron oxides, however, will absorb ultra-violet radiation and thus have a protective effect in paints for exterior use.

Cadmium yellows, CdS: synthetic pigments varying in colour from primrose to orange, the shade being determined by the process conditions. Stable (up to about 500°C) and non-toxic, cadmium yellows are lightfast and unaffected by alkalis although they are decomposed by acids. They may be used on a variety of substrates, especially where alkali resistance and/or temperature resistance, e.g. in stoving finishes, is required.

Green inorganic pigments

Chromium oxide, Cr_2O_3: dull green, synthetic pigment having good stability to light, heat, acids and alkalis but of poor opacity. Useful as a pigment for paints required to exhibit high chemical and photochemical resistance, chromium oxide is also used as a colorant for cement and concrete.

Lead chrome greens, $PbCrO_4$:$KFe(Fe(CN)_6)$ (Brunswick greens): synthetic pigments, ranging from grass green to deep green in colour, the shade being controlled by the Prussian blue ($KFe(Fe(CN)_6)$) content admixed with lead chrome yellow. While the pigments have good opacity, they tend to darken in polluted atmospheres and are affected by alkalis. They are also subject to floating or flooding in certain paint media.

Blue inorganic pigments

Prussian blue, $KFe(Fe(CN)_6)$, potassium ferric ferrocyanide: bright blue synthetic pigment having high staining power but of low opacity. It is lightfast and reasonably acid-resistant, but Prussian blue is decomposed to iron oxide by alkalis and by elevated temperatures.

Ultramarine blue, $3Na_2O_3.3Al_2O_3.6SiO_2.2Na_2S$: complex synthetic aluminosilicate pigment; it occurs naturally as the semi-precious stone lapis-lazuli. The colour of the pigment is ascribed to the presence of sodium sulphide, although the shade can be altered by varying the amount of silica raw material used in its manufacture. Of coarse texture and subject to hard settlement, the pigment has good lightfastness, heat resistance and alkali resistance, but it is decomposed by acids. Although it has a poor tinting strength, ultramarine blue can be used as a pigment for paints intended for alkaline substrates and as a colorant for cement and concrete.

Red inorganic pigments

Red iron oxide, Fe_2O_3(haematite): both natural and synthetic red iron oxides are used as pigments; the former, mined as the ore haematite, are normally calcined. The presence of impurities together with variation in the Fe_2O_3 content yields pigments of a wide range of shades from orange-red to deep brown. The Fe_2O_3 content also affects the opacifying properties of the pigments, opacity increasing with the oxide content. The natural oxides are used as colorants in timber stains and for cement and concrete as well as in low-cost metal primers.

The synthetic oxides have a softer texture and greater purity, the latter giving brighter, cleaner colours and greater tinting strength. The colours are determined by the manufacturing conditions and the improved colour properties of the synthetic oxides permits them to be used in a wider variety of paint systems.

Both the synthetic and natural oxides are resistant to alkalis and organic acids, but they are affected by mineral acids and elevated temperatures. Furthermore, they will absorb ultra-violet radiation and hence increase the resistance of paint films to photochemical degradation.

Two other inorganic red pigments are also known, namely red lead and basic lead silicochromate. Both are primarily used for their corrosion-inhibiting action (see pages 20 and 21).

Black pigments

Black iron oxide, Fe_3O_4(magnetite): a synthetic pigment of low tinting strength, mainly used as a colorant for fillers, primers and undercoats. The pigment colour is reddened by heat but otherwise the oxide has a high chemical resistance.

Carbon blacks: pigments produced by carbonisation of hydrocarbon minerals and vegetable and animal matter; the source and particularly impurities affect the properties. Purer carbons (i.e. a high carbon content) are of fine texture, pure colour and good opacity, these properties being degraded with lower carbon contents.

Carbon blacks are lightfast, insoluble in acids and alkalis and are resistant to solvents. They have high opacities, even at low addition levels, and high tint strengths. They may be used as colorants in a wide variety of paint systems.

Yellow organic pigments
Hansa yellows: azo dyes, ranging in colour from orange to green-shade yellows, characterised by good light-stability in full shade although this is lost on reduction with white pigments. While they are of low opacity, their absence of toxicity renders them more suitable than lead chromates in many applications. Hansa yellows are soluble in ketones, esters and aromatic hydrocarbons but poorly soluble in aliphatic hydrocarbons. This renders them suitable for use in air-drying coatings containing aliphatic hydrocarbons and in emulsion paints.

Benzidine yellows: insoluble azo dyes prepared in a range of shades, from yellow to red. They are of greater opacity than the Hansa yellows and are insoluble in most solvents used in paint formulations. They are non-toxic, with good resistance to acids and alkalis as well as to temperatures up to about 140°C. However, benzidine yellows are subject to degradation by light, even when used at full shade. While they are suitable for interior use, particularly for stoving finishes and as replacements for lead chrome pigments, their use for exterior coatings is limited.

Green organic pigments
Pigment green B: a toner which, unlike the azo dyestuffs, is not purely organic in nature but is a saltlike compound. (Toners are formed by reaction of a basic dyestuff with a complex acid or the reaction of an acidic dyestuff with a metal or metal salt; they are generally more resistant to solvents than the organic dyes.) Pigment green B has a blue-shade green colour of good opacity in paints. It has good resistance to solvents, heat and alkalis, but is decomposed by acids. It is widely used in decorative finishes, particularly if alkali resistance is required, and for stoving-cured coatings; it is also used as a colorant for cement and concrete.

Blue organic pigments
Phthalocyanine blues: varying in colour from reddish-blue to yellowish-green, characterised by high tint-strengths and opacities combined with good lightfastness. They are non-toxic, temperature-stable (up to about 500°C) and resistant to most solvents and chemicals except strong acids. Phthalocyanine pigments are used in a wide variety of paint coatings because of their high chemical resistance and lightfastness.

Red organic pigments
Toluidine reds (Helio reds): non-toxic azo dyes characterised by bright, clean colours, good lightfastness and high opacities although the colour

stability may be decreased when they are reduced with white pigments. They are produced by coupling of diazotised *m*-nitro-*p*-toluidine with β-naphthol, the colour being determined by the reactants and reaction conditions.

Toluidine reds have good colour stability and resistance to acids and alkalis, and to heat up to about 180°C for short periods. They are soluble in aromatic solvents and slightly soluble in aliphatic solvents and alcohols, so that they are susceptible to bleeding in many non-convertible and stoving systems.

Arylamide reds: non-toxic azo dyes ranging in colour from orange through red to deep crimson. They are characterised by good lightfastness, even when greatly reduced with white pigments, and are temperature-stable (up to about 140°C). Arylamide reds are slightly soluble in solvents. They are resistant to acids and alkalis and because of their colour-stability they can be used in decorative exterior coatings at all levels of reduction.

1.3.3 CORROSION-INHIBITORY PIGMENTS

Pigments used to protect metallic substrates against corrosion fall into two broad classes, metals and salts. Of the former, lead and zinc are the most important, although powdered aluminium and stainless steel have some application. Corrosion-inhibitory salts contain water-leachable anions which can passivate the metal or affect the corrosion process. With these pigments, the mechanism of protection differs from that of metals.

Inorganic salts

Various inorganic salts function as corrosion-inhibitors when incorporated in paints, usually primer coats. Until recently, lead- and chromate-containing compounds were the main inhibitory pigments but the toxicity and pollution problems associated with such materials has resulted in a tendency to use alternative pigments. Although newer inhibitive pigments such as zinc phosphate are less effective than some of the traditional materials, they still are able to confer considerable protection upon the substrate.

Basic lead silicochromate, $PbO.\ CrO_3.SiO_2$: a dull-orange synthetic pigment of low tint-strength, permitting it to be used in admixture with colouring pigments at all levels in the coating system to provide 'in depth' corrosion protection. The protective action appears to derive from both the formation of inhibitive lead soaps (in paints of high acid value) and leaching of chromate ions. It can be used in a wide variety of paint media without deleterious effects on the stability or flow properties of the liquid paint. Although it contains lead, lead silicochromate is somewhat less toxic than the traditional red lead pigment.

Basic lead sulphate, $2PbSO_4.PbO$: a white synthetic pigment of moderate tint-strength which does not adversely affect the flow properties of the liquid paint. Consequently, basic lead sulphate can be used at all levels in a coating system. However it darkens in polluted atmospheres, and this, together with the toxicity of the pigment, limits its use in many applications. Its primary use is in anti-corrosion coatings for steel, particularly in marine environments, and it is commonly used in association with aluminium and/or micaceous iron oxide. The protective action of the pigment is usually ascribed to polarisation of the electrochemical reactions involved in the corrosion process.

Calcium plumbate, $2CaO.PbO_2$: a synthetic pigment, white when pure but usually buff-coloured. Being basic in nature, calcium plumbate reacts with acidic paints to yield lead soaps which have an inhibitory action as well as a reinforcing effect on the mechanical properties of the film. The pigment has no detrimental effect on the liquid paint stability provided the acidity is not too high. The protective action of calcium plumbate is associated with polarisation of the electrode reactions. This pigment is especially suited for use in primers for galvanised steel, although its toxicity does limit its applications in modern technology.

Red lead, $PbO_2.2PbO$: basic, orange-coloured synthetic pigments mainly used as corrosion-inhibiting rather than colouring pigments, usually in combination with linseed oil. Being basic, red lead reacts with acidic paints to form lead soaps which have a protective effect as well as reinforcing the paint film. However, the liquid paint will progressively thicken as a result of reaction between red lead and the binder acid. Although red lead–linseed oil combinations are perhaps the most effective protective primers for steelwork, the toxic nature of the pigment now limits its use in modern paint technology.

Zinc (potassium) chromate, $K_2Cr_4.3ZnCrO_4.Zn(OH)_2$: zinc chromate used in protective primers is required to be free from residual chloride, in contrast to the pigment used solely for colouring, and a different preparative route is often used. The anti-corrosion action of the pigment is apparently due to the presence of leachable chromate ions, although it has little effect in chloride-containing environments. Zinc chromate is basic in nature and this limits its use in acidic paint media.

Zinc phosphate, $Zn_3(PO_4)_2$: a non-basic, white synthetic pigment which can be used in almost all paint media. Being virtually insoluble, zinc phosphate does not leach from paints as do more soluble salts such as zinc chromate. The mechanism of the protective action of zinc phosphate is uncertain: it appears to polarise both the anodic and the cathodic corrosion reactions.

Zinc tetroxychromate, $ZnCrO_4.4Zn(OH)_2$: a yellow synthetic pigment exclusively used in metal pretreatment primers (also known as wash or etch primers). Less soluble than zinc chromate, it has a protective action as well as functioning as an adhesion promoter for subsequent coats applied over the pretreatment primer when applied to light alloys and steelwork. The protective action of the pigment appears to be due to slow release of chromate ions.

Other protective pigments: several novel inhibitive pigments have recently been studied for their applicability in anti-corrosion coatings. These pigments include calcium chromate, calcium molybdate, magnesium phosphate, chromium phosphate, zinc molybdate, barium metaborate, barium chromate, zinc phospho-oxide complex and organonitrogen compounds. While many of these pigments can be used as the sole protective agent in primers, they appear to be more effective when used in conjunction with one or more of the newer and/or traditional inhibitive pigments.

At present the above and other new inhibitive pigments are still under investigation for their potential uses in paints and are far from being widely employed. Nevertheless, they are mentioned here since most of them are non-toxic and this is an important consideration in view of the increasing attention being devoted to health and safety and to ecology.

It is perhaps of interest to note here that there is now some evidence to suggest that the extender talc (see page 24) also functions as a corrosion inhibitor. Other extenders such as mica have also been reported to improve the corrosion protection of inhibitive pigments in various media.

Metallic pigments

Four metals are used as pigments in paints: aluminium, stainless steel, lead and zinc. Of these, only lead and zinc have a definite anti-corrosive action.

Stainless steel: an alloy of iron, chromium and nickel with various minor alloying elements. Massive stainless steel is a hard but ductile corrosion-resistant metal, the corrosion resistance being due to the presence of a complex oxide surface film. When used as a pigment, particulate stainless steel has been claimed to provide corrosion protection although the mechanism for this is uncertain. Presumably the particles constitute a barrier against the ingress of corrosive anions, although it is conceivable that a more complex electrochemical process is involved.

A stainless steel pigment may also provide a decorative effect, particularly as the metal itself is essentially non-reactive in paint media.

Aluminium: a pure metal that is rendered resistant to many aggressive agents by the presence of a surface film of alumina. It is available, when powdered, in two forms—the so-called leafing and non-leafing grades. Particles of the leafing grades, which are usually surface-coated with stearic acid or similar surface-active agent, tend to float towards the surface of a paint film and become oriented parallel to the plane of the film. The close-packed layer of aluminium platelets constitutes in the film a reasonably efficient moisture barrier and thus functions as a protective pigment. Normally, relatively low additions levels are used, often without the incorporation of other pigments. It is possible that the aluminium particles also function as sacrificial anodes if their surface oxide film is disrupted, for example, as the result of mechanical damage to the paint.

The non-leafing grades do not migrate to the film surface, but are more randomly distributed within the film. Again, low addition levels are used and the main function of non-leafing aluminium pigmentation is to provide a decorative (sparkle) effect to the coating.

Lead: a heavy, malleable metal that is resistant to a wide variety of chemicals as a result of the formation of protective surface oxide and salt films. In paint technology, it is mainly used for the preparation of protective primers, often based on linseed oil, for steelwork. High addition levels are used and, because of the high density of lead, settlement can be a problem during storage of the liquid paint.

The mechanism of protection of lead primers is complex but may involve both the formation of lead soaps, which may block permeation pathways, and a quasi-sacrificial anode effect, particularly in aggressive environments.

Zinc: an electropositive metal that is chemically reactive. It is mainly used in paint technology for the preparation of zinc-rich protective primers for steelwork. The zinc particles function as sacrificial anodes (i.e. cathodically protect the steel) and thus very high addition levels are required to ensure the inter-particle and particle–substrate contact necessary for electrical conduction. Zinc-rich primers function as efficient anti-corrosion coatings but the high level of zinc addition can lead to settlement problems during storage of the liquid paint.

1.4 Extenders

Extenders are paint additives that are insoluble in the binder (and solvents) and which have little or no opacifying or colouring effect on the film. They are added to modify the flow properties of the liquid paint as well as the mechanical, permeability, gloss and levelling characteristics of the paint film.

Barytes, $BaSO_4$: a hard, dense, naturally occurring inorganic mineral which is very resistant to the action of acids and alkalis. A synthetic form of barium sulphate, known as blanc fixe, is also used as an extender and has similar properties to barytes. It is transparent in oil and resinous media so that it can be incorporated in paints based on solvent-soluble binders without adversely affecting the colour or opacity. However, the high specific gravity of barytes can result in settlement problems during paint storage. It is used as a chemically inert filler which also reinforces the mechanical properties of the film; however, when used at high levels, barytes (in common with certain other extenders) can reduce the film durability.

China clay, $Al_2O_3.2SiO_2.2H_2O$ *(kaolin):* a naturally occurring hydrated aluminium silicate used in a fine particulate form. It is used as a filler in solvent-containing paints but at low addition levels, since the small particle size can adversely affect the flow characteristics of the liquid paint. The fine particles, however, will reduce settling tendencies in paints during storage. China clay has a 'flatting' (gloss-reducing) effect and is used as a flatting agent in undercoats and eggshell-finish paints. China clay is also widely used in emulsion paint formulations.

Mica, $K_2O.2Al_2O_3.6SiO_2.2H_2O$*:* a naturally occurring potassium alumino-silicate used in the form of thin platelets. The lamellar structure of mica results in a leafing effect, like the leafing grades of aluminium, that markedly reduces water permeability in the paint film. The incorporation of mica in a paint improves weathering resistance by reducing the tendency to fail by cracking and checking. This property, combined with the reduced permeability of mica-containing paints, permits mica to be used in exterior coatings, especially where protection as well as decoration is required.

Talc, $MgH_2(SiO_3)_4$ *(French chalk):* a naturally occurring hydrated magnesium silicate mineral. It occurs as a mixture of lamellar and fibrous particles, the latter providing a reinforcing effect while improving film flexibility. The platelets have an effect akin to that of mica in reducing water permeation. Consequently, talc is often used in protective paints when high durability is required, and it has similar applications to mica. Both mica and talc are thought to have a beneficial effect on corrosion resistance when incorporated in primers and finishing coats. This effect may be due to the reduced water permeability or to some more complex phenomenon.

Whiting, $CaCO_3$ *(chalk, Paris white):* a mineral occurring naturally as limestone as well as being produced synthetically. It is widely used in interior and exterior emulsion paints and in undercoats but, since whiting is attacked by mineral acids, it is not used in aggressive environments. Nevertheless, it has wide applicability in a range of coating systems, both aqueous- and solvent-based.

1.5 Solvents

Solvents are volatile liquids added to paints to dissolve the binder (the resin component) and/or to modify the paint viscosity. Thus not only must a solvent dissolve the resin but the resulting solution should have a viscosity that is in accord with the proposed storage and application modes of the

Table 1.4 Solubility Parameters for Classes of Solvent

Solvent class	Hydrogen bonding		
	Strong, δ_s	**Moderate, δ_m**	**Poor, δ_p**
Alcohols	11–13	–	–
Ketones	–	8–10	–
Ethers	–	9–10	–
Esters	–	8–9	–
Aliphatic hydrocarbons	–	–	7–8
Aromatic hydrocarbons	–	–	8–9

liquid paint. Consequently, the solvent must have an evaporation rate that permits the paint to deposit a film possessing the required characteristics. Furthermore, ideally, the solvent should be non-toxic, of low cost and have an acceptable odour.

The above selection criteria are valid for solvents used for convertible binders. In non-convertible coatings, solvents have a more complex function in that they determine, wholly or in part, the coating application characteristics as well as the drying time and the properties of the final film. Consequently, blends of two or more solvents are used, the blend comprising a solvent for the binder together with a diluent. The diluent may not be a true solvent for the binder but will aid its dissolution in the primary (or true) solvent; it is, however, of lower cost than the true solvent so that its inclusion reduces the overall cost of the formulation.

The selection of a solvent was formerly made on the basis of 'like dissolving like' but now involves the concept of solubility parameters. There are three classes of solubility parameter, namely strongly hydrogen bonded (δ_s), moderately hydrogen bonded (δ_m) and poorly hydrogen bonded (δ_p). Alcohols fit into the first category, ketones, ethers and esters into the second, while hydrocarbons comprise the third class. The approximate ranges of these parameters for the various classes of solvent are given in Table 1.4. Resins used as paint binders have a range of parameter values (Table 1.5) and solvent selection is based on whether the resin parameters coincide with those of the proposed solvent. Thus epoxy resins, which have a δ_m range of 8–13, are soluble in ketones, ethers and esters but insoluble in hydrocarbons and alcohols because the δ_p and δ_s values are zero.

Conversely, solvent resistance in a coating is achieved by using as the binder a resin having a solubility parameter (δ) significantly different from that of the solvent in question. Compatibility of two or more resins in a coating requires that there is normally no more than 1 unit difference in the midpoints of their δ ranges.

Table 1.5 Solubility Parameter Ranges of Resins

	Solubility parameters		
	δ_s	δ_m	δ_p
Alkyds			
Short oil	9–11	7–12	8–11
Medium oil	9–11	7–12	7–11
Long oil	9–11	7–10	7–11
Vinyls			
Poly(vinyl chloride–vinyl acetate)	0	7–14	9–11
Poly(vinyl chloride–vinyl acetate–dibasic acid)	0	7–12	10–11
Poly(vinyl chloride)	0	9–10	10–11
Poly(vinyl butyral)	9–15	9–11	0
Epoxies			
Epoxy equivalent:			
400–500	0	8–13	10–11
800–900	0	8–13	0
1700–2000	0	8–13	0
2000–4000	0	8–10	0
Drying-oil epoxy	0	7–10	8–11
Others			
Poly(methylmethacrylate)	0	8–13	8–13
Polyurethane	0	8–12	8–11
Polyester	9–11	7–12	8–11
Chlorinated rubber	0	7–11	8–11

Usually, when selecting solvents for a particular resin, the solubility parameters of the solvents have to lie within the range of the resin; other selection criteria include cost, paint viscosity and drying time. The amount of non-solvent or diluent that can be blended with a solvent is normally that which for the mixture gives a mean δ that lies within the mid-80 per cent of the δ range of the resin.

The evaporation rate of a solvent is particularly important for non-convertible coatings. Slow-evaporating solvents may give unduly long drying times which could give rise to problems in batch coating operations or excessive flow and sag of the wet film (additives to offset such effects are discussed on pages 32 and 48). High evaporation rates may adversely affect flow and film integrity and probably constitute a greater problem. Selection is based on the mode of coating application and, typically, slower evaporation rates are required for brush-applied coatings while faster rates are used for spray methods. Various properties of some common solvents are given in Table 1.6.

Table 1.6 Solvent Properties

Solvent	Specific gravity	Boiling-point (°C)	Evaporation Rate*	Flash-point (°C)
Acetone	0·79	56	944	−18
Butyl acetate	0·88	125	100	23
Butyl alcohol	0·81	118	36	35
Ethyl acetate	0·90	77	480	−4·4
Ethyl alcohol	0·79	79	253	12
Ethylene glycol monoethyl ether (2 ethoxyethanol)	0·93	135	24	49
Methyl ethyl ketone	0·81	80	572	−7
Methyl isobutyl ketone	0·83	116	164	13
Toluene	0·87	111	214	4·4
White spirit	0·80	150–200	~18	38(min.)
Xylene	0·87	138–144	73	17–25

*Evaporation rate values based on butyl acetate = 100; these values relate to pure solvents and will be lower where dissolved solids or diluents are present.

1.5.1 HYDROCARBON SOLVENTS

Toluene, $C_6H_5.CH_3$: an aromatic hydrocarbon used in solvent blends for air-drying vinyl and chlorinated rubber coatings as well as a diluent in nitrocellulose paints.

White spirit: largely aliphatic (but containing 15–18 per cent aromatics) blend of paraffins of variable composition. Slow-evaporating and dissolving most natural resins, oleoresinous varnishes and medium- and long-oil alkyd resins, white spirit is commonly used as the main (often sole) solvent for brush-applied decorative and protective paints based on these binders. It is also used as a cleaning and degreasing solvent.

Xylene, $C_6H_4(CH_3)_2$: an aromatic hydrocarbon widely used as a solvent for short-oil alkyd, vinyl, chlorinated rubber and polyurethane resins. The good solvating power and moderate evaporation rate permit xylene to be used for sag-resistant spray-applied coatings (even of relatively thick films) as well as for stoving-cured coatings.

1.5.2 ALCOHOLS AND ETHERS

Butyl alcohol, C_4H_9OH (butanol): slow-evaporating solvent for a wide variety of oils and resins, particularly amino and acrylic resins, and also in solvent combinations for nitrocellulose resins.

Ethyl alcohol, C_2H_5OH (ethanol): usually used admixed with methyl alcohol, dyes and toxins in the form of industrial alcohol (methylated spirits). Ethyl alcohol, a fast-evaporating solvent, is used for poly(vinyl butyral) as well as with other solvents for nitrocellulose.

Ethylene glycol monoethyl ether, $C_2H_5.O.CH_2CH_2OH$ (2-ethoxyethanol): a slow-evaporating solvent for many resins which is often added to brush-applied formulations that cannot incorporate aliphatic hydrocarbons such as white spirit. The high-boiling ethers are widely used as coalescing aids (solvents) in emulsion paints, ethylene glycol monobutyl ether acetate being a typical example.

1.5.3 ESTERS AND KETONES

Acetone, $CH_3.CO.CH_3$ (dimethyl ketone): fast-evaporating powerful solvent used for vinyl copolymers and nitrocellulose. It is also blended (at low addition levels) with many other solvents, when its high solvating power and evaporation rate modify the properties of the liquid paint and the film.

Methyl ethyl ketone, $CH_3.CO.C_2H_5$ (MEK): powerful solvent with a fast evaporation rate, widely used for vinyl copolymers, epoxy and polyurethane systems. It is often blended with less powerful solvents to modify the film-forming properties and application characteristics of coatings.

Methyl isobutyl ketone, $CH_3.CO.CH_2CH(CH_3)_2$ (MIBK): a fast-evaporating powerful solvent used in similar applications to MEK but where a somewhat slower evaporation rate is required. It is also used at low addition levels in solvent blends where its fast evaporation rate and high solvent power will benefit the properties of the liquid paint and film-forming characteristics.

Butyl acetate, $CH_3.COO.C_4H_9$: ester solvent with moderately fast evaporation rate, of general applicability. Formerly butyl acetate was the major solvent for nitrocellulose coatings; it is now used for a wide variety of synthetic resins but is less powerful than the ketone solvents.

Ethyl acetate, $CH_3.COO.C_2H_5$: ester solvent with rapid evaporation rate and a pleasant fruity odour which is often preferred to that of the more pungent ketones. The major application of ethyl acetate was formerly in nitrocellulose coatings but it now has a somewhat wider application. It has a lower solvating power than the ketone solvents.

1.6 Plasticisers

Plasticisers are organic additives that are used to increase the film flexibility of paints, particularly those based on binders that otherwise would yield brittle films. Plasticisation can be achieved internally or externally. Internal plasticisation consists of copolymerisation of the brittle resin, the major binder component, with a second resin in order to modify the properties, e.g. vinyl acetate will 'plasticise' the rather rigid and brittle vinyl chloride. This approach is effective but somewhat limited in scope and therefore external plasticisation is more widely adopted. External plasticisation usually consists of the incorporation of a compatible low-molecular-weight non-volatile liquid with the resin, although resins can also be plasticised with high-molecular-weight resins, as in the addition of alkyd resins to chlorinated rubber.

The plasticiser must be non-toxic and should provide for maximum film flexibility while having a minimal softening effect and colour distortion, partcularly on external exposure. The amount and type of plasticiser added varies with the resin type as well as the proposed application of the coating.

A plasticiser functions by reducing the glass transition temperature (T_g) of the resin, the T_g being the temperature at which a resin changes from a hard, brittle solid to a rather rubbery state. Two classes of plasticiser are known: (i) solvent or primary and (ii) secondary (non-solvating).

Primary plasticisers act as solvents for the resin by virtue of chemical groups which can interact with the polymer and so render it compatible. Thus the plasticiser will become incorporated into the structure and reduce the rigidity although the mechanical properties of the film may be impaired.

Secondary plasticisers have no solvent action, are non-reactive and may only be compatible over a limited addition range. These plasticisers only have a mechanical (lubricating) effect and, consequently, reduce the film strength to a lesser degree. However, it is possible for secondary plasticisers to migrate or be leached from the film, leading to loss of flexibility.

Plasticisers affect many properties of the paint film, including tensile strength, toughness, extensibility, permeability and adhesion. All of these properties are affected to different degrees, often in a complex manner, by plasticisation but the detailed effects vary with the polymer and plasticiser types. In general, the plasticiser reduces the tensile strength while increasing the extensibility. Permeability may remain largely unchanged up to a certain critical plasticiser level, thereafter it can increase sharply. The film toughness and adhesion normally increase with the plasticiser level up to a peak value, thereafter decreasing again. The desired effects on the mechanical properties of the film may adversely affect other properties and this consideration often dictates the optimum level of plasticiser that may be added to the formulation.

Chlorinated paraffins: used primarily for chlorinated rubbers without adversely affecting the chemical resistance of the film. Very high addition levels (up to about 50 per cent) are used.

Dibutyl phthalate (DBP): extensively used plasticiser of good compatibility with a wide range of resins. Moderately resistant to yellowing, DBP has, however, a degree of volatility and this limits its applicability. It is widely used for nitrocellulose, at levels of 20–50 per cent, and poly(vinyl acetate) emulsion coatings, at levels of 10–20 per cent (added during emulsion polymerisation).

Dioctyl phthalate (DOP): less volatile and more resistant to light and heat than dibutyl phthalate. It is widely used for nitrocellulose and for poly(vinyl chloride) organosol and plastisol coatings.

1.7 Driers

Driers, or drying accelerators, are oil-soluble metal salts of organic acids added to oxidation-curing systems to increase the curing (drying) rate of the applied film. The most widely used salts are the naphthenates, octoates, rosinates and linoleates of lead, cobalt and manganese, although a variety of other organic acids, as well as metals such as aluminium and zirconium, are also employed. The use of other driers such as zirconium salts is increasing because of concern over the toxicity of lead compounds.

Driers increase the curing rate by accelerating the oxidation of the unsaturated oil component of the binder; the exact mechanism for this is unknown. The most effective driers are lead and cobalt salts, which have different modes of drying action. They are usually used in combination, a lead to cobalt ratio of 10:1 being typical. Lead salts are through-film (or within-film) driers, while cobalt salts are surface driers, so that a combination of both yields a uniform drying rate. With alkyd binders, the combination of driers will reduce the bloom tendency of gloss paints; oil-soluble calcium salts are also sometimes added to reduce bloom although they have no drying action.

The acid radical of the drier has no effect on the drying rate but can influence the film properties. Naphthenates have no deleterious effects but rosinate driers can cause accelerated film embrittlement, while rosinate and linoleate driers tend to lose efficacy upon prolonged storage.

The drier content of a formulation is somewhat critical if film rivelling (due to uneven drying), bloom, etc., are to be avoided. Typical addition levels are 0·25–0·5 per cent lead metal and 0·025–0·05 per cent cobalt metal of the solids content of the binder.

1.8 Bioactive additives

In marine environments, especially when submerged, surfaces tend to ‘foul’, that is, accumulate deposits and growths of marine organisms such as weed, tube worms, barnacles and bacteria. Fouling, particularly when heavy, increases the frictional drag on ships and also accelerates corrosion, and necessitates expensive periodic dry docking for its removal. To prevent this, semi-permanent anti-fouling compositions are applied to the surface, these coatings being applied as a sub-layer beneath the coating system. They contain toxins that are poisonous to marine fouling organisms and which are progressively leached out at a rate sufficient to inhibit attachment and proliferation of organisms. The commonly used anti-fouling toxins are mercuric oxide, tributyl tin oxide, metallic copper and copper suboxide. High addition levels are used to ensure a large reservoir within the coating system.

Paints, particularly water-based emulsions, are susceptible to attack by micro-organisms such as fungi and bacteria. Apart from the obvious impairment of appearance, paint coatings may suffer degradation by micro-organisms. Bioactive additives are required to prevent this attack.

Biocides or anti-bacterial agents are incorporated within emulsion paints to prevent bacterial (and fungal) attack on the thickener, usually a cellulose ether. Biocides function by interfering with the bacterium's metabolism, thereby preventing enzyme synthesis, enzymes (rather than the bacteria themselves) being the active agent in paint degradation. The commonly used biocides are complex phenol or formaldehyde compounds and sometimes organomercurials, usually at a level of 0·05–0·3 per cent by weight of the paint. It is important, however, to ensure during paint manufacture that the raw materials do not contain enzymes, otherwise degradation will still occur despite the incorporation of a biocide, since the latter has no toxic effect on enzymes.

Fungal growth will contribute to the degradation of a paint coating but more commonly it impairs its appearance. In certain circumstances there may be an associated health hazard, as in kitchens, dairies and breweries. Fungicides are toxic agents that are included in paint formulations to inhibit the colonisation and proliferation of fungi. A range of toxins is used for this purpose.

Barium metaborate: good fungicidal action provided it is used at levels of 15–20 per cent. It is slightly water soluble and therefore best used in media of low permeability.

Dichlorofluamide: broad-spectrum fungicide, used at levels of 1·5–2·0 per cent, that is unstable in water-based paints because of hydrolytic breakdown in their somewhat alkaline conditions.

Dithiocarbamates: broad-spectrum fungicides of low solubility used at addition levels of 3–6 per cent of total solids. They are not generally used in oxidative-cured paints because they markedly tend to reduce the drying time.

Organomercurials: highly toxic (to humans) but effective fungicides that are generally broad-spectrum in their activity, e.g. phenyl mercury acetate. They are limited in applicability, however, because of their leachability, tendency to darken in sulphur-containing atmospheres and susceptibility to photochemical degradation.

Organotin compounds: of lower toxicity to humans than organomercurials but fairly effective, particularly for interior coatings. Typical addition levels are 1 per cent of total solids.

Zinc oxide: traditional fungicide that also has a pigmentary effect (see page 16). High addition levels (about 30–40 per cent) are necessary. It is less used in modern formulations because of its poor exterior durability and rather variable efficacy.

1.9 Viscosity modifiers and anti-settling agents

Paint formulations may yield liquid paints that are too fluid. With a liquid of low viscosity the flow may be excessive for some purposes, although high flow may be advantageous in some circumstances. Thus low viscosity is desirable in stains, sealers and spraying lacquers, but not in high-build coatings and brush-applied paints. Settlement of pigments in storage, particularly if they are dense, occurs readily in low-viscosity paints, and heavy settlement can cause problems in redispersing the pigment. These problems of flow and pigment settlement can be offset by adjusting the paint viscosity. In the case of pigments, further adjustments can be made to the formulation to counteract settlement.

Viscosity can be increased by means of thickening agents which increase the viscosity (i.e. decrease flow) without introducing thixotropy. Thixotropic agents provide the paint with a gel-like structure at zero or low shear rates, but flow will occur at high shear stresses. A gel state is resumed once shear stresses are reduced to zero. Thixotropy is a useful attribute for many brush-applied paints in that sagging and runs are obviated. This property will also reduce, or even entirely eliminate, settlement during storage.

Thixotropic paints are not desirable in certain applications, for example where good flow is required, and in such situations settlement of pigments can be reduced by incorporating anti-settling or surface-active agents such as soya lecithin, usually at a level of about 1 per cent on the formulation. The surface-active agent is adsorbed onto the pigment surface, increasing its volume and decreasing the density. The overall effect is a reduced settling tendency. Redispersion of settled pigments is facilitated by incorporating surface-treated extenders, typically calcium carbonate at a level of 5 per cent of the pigment content. These surface-treated extender particles have a large volume due to the organic surface layer, and during settlement they pack between the pigment particles. When the paint is stirred, the extender particles aid breakdown of pigment agglomerates so that redispersion is more easily achieved.

Many compounds will increase paint viscosity and/or introduce thixotropy into the formulation. The more common additives are described below.

Cellulose ethers: modified cellulose ethers, i.e. sodium carboxy methyl cellulose and ethyl hydroxy ethyl cellulose, are widely used as thickening agents for emulsion paints. The compounds imbibe water and swell but at their normal addition level, i.e. 0·25–1·0 per cent on the paint weight, they do not impart thixotropy to the formulation. Sodium carboxy methyl cellulose gives superior flow and levelling characteristics to the formulation, while the ethyl hydroxy ethyl derivative gives better wash and scrub resistance.

Micronised silica: an inert, microfine form of silica which absorbs water and other fluids to form stable gels. The chemical inertness of silica renders it suitable for thickening and gel-forming with a wide variety of media. The silica can be surface-treated to make it compatible with non-aqueous and non-polar fluids.

Natural clays (bentonites): these naturally occurring colloidal silicate clays swell when dispersed in water and so are used as thickening agents for water-based paints. If they are amine-treated, bentonite clays will thicken and confer thixotropy upon solvent-soluble paints, addition levels of 2–4 per cent of the paint weight being used. Thixotropic gels are prepared by dispersing the amine-treated bentonite in a hydrocarbon solvent which is admixed with the paint; the addition of a small amount of polar solvent causes gel formation.

1.10 Pigment-dispersing agents

Paint formulations obviously require an even dispersion of pigments and extenders throughout the liquid medium. This involves, during paint manufacture, dispersion of the pigments and extenders in the liquid medium to break down agglomerates and uniformly wet all the particles. For this to occur, the medium must displace any moisture, occluded gases and fine debris on the particle surfaces: how readily this process is effected is thus of practical importance. While most (but not all) pigments and extenders are oleophilic, that is, they are wetted by solvent-soluble media, only a proportion are hydrophilic or wetted by water: for example, many organic pigments are hydrophobic or water-repellent.

Pigment-dispersing agents, or surface-active agents, are used to facilitate pigment and extender dispersion where wetting by the medium is poor or even non-existent. They are often used routinely even when there is good wetting, to ensure good dispersion. The various surface-active agents used for this purpose include sulphonated oils, sodium or potassium salts of carboxylic (fatty) acids, quaternary ammonium compounds and condensation products of ethylene oxide with fatty acids or amides. Addition levels are usually of the order of 0·1–0·5 per cent of total paint solids.

These surface-active or dispersing agents are compounds having both polar (hydrophilic) and non-polar (oleophilic) groups in their structures. When applied to the pigment or extender particles, either the polar or the non-polar portion adsorbs onto the surface with the other, 'non-reacted', portion being oriented away from the surface; the group that attaches obviously is determined by the particle in question, i.e. a hydrophobic surface will adsorb the oleophilic portion and vice versa. This adsorption process displaces moisture, gases, etc., from the particle surface.

In the case of a hydrophilic pigment, which would not be wetted by the paint medium, the oleophilic portion of the surface-active agent interacts with the solvent or oil medium so that there is readier wetting of the particles. As a result, dispersion of the pigment and/or extender is facilitated.

1.11 Bibliography

Tawn, A.H. (Ed.), *Paint Technology Manual,* Part 1, *Non-convertible Coatings,* 2nd edn. Chapman and Hall: London, 1969.

Bews, I.C.R. (Ed.), *Paint Technology Manual,* Part 3, *Convertible Coatings.* Chapman and Hall: London, 1966.

Newton, D.S. (Ed.), *Paint Technology Manual,* Part 2, *Solvents, Oils, Resins and Driers,* 2nd edn. Chapman and Hall: London, 1969.

Tawn, A.H. (Ed.), *Paint Technology Manual,* Part 6, *Pigments, Dyestuffs and Lakes.* Chapman and Hall: London, 1966.

Patton, T.C., *Pigment Handbook,* vols I, II and III. Wiley: New York, 1973.

Roff, W.J. and Scott, J.R., *Fibres, Films, Plastics and Rubbers—Handbook of Common Polymers.* Butterworth: London, 1971.

Saunders, K.J., *Organic Polymer Chemistry.* Chapman and Hall: London, 1973.

CHAPTER 2

Essential Concepts of Paint Formulation

2.1 Introduction

The underlying principles governing the formulation of various paint types, and the properties of the major components used, are described in other chapters. Likewise, some of the more important mathematical techniques of formulation, manipulation and statistical interpretation of data will be covered in Chapter 3. In this chapter the technique of paint formulation is discussed, that is the processes and theory involved in combining the various components to obtain a paint that satisfies the specification requirements.

Most paints are composed of pigments and possibly extenders, a binder or combination of binders, and a solvent blend as well as several minor additives. There are of course exceptions, notably clear lacquers and varnishes, which contain no pigment or other particulate material, and solventless coatings based entirely upon either low-viscosity liquid resin systems or powdered combinations of resins in which, after application, flow is induced by a stoving process.

The absolute amounts and relative proportions of the various ingredients play a major role in determining the performance of the resultant coating formulations although, obviously, the characteristics of the selected components must also be suitable for the coating in its service environment. In fact, the incorrect choice of only one component can reduce a paint's performance even if the other additions and general formulation requirements are otherwise satisfactory. This is a frequent cause of premature paint failure in practice.

Given that the constituents selected are suitable for the end use, then the most fundamentally important characteristic of a paint is its pigment volume concentration, that is the ratio of the volume of pigments and extenders to that of the non-volatile binder content of the formulation. Other formulation aspects must also be considered: for instance, the solvent composition affects such properties as drying time, film formation and application characteristics. It also plays a major role in controlling and modifying the viscosity and flow characteristics of the resultant coating. These parameters, pigment loading, solvent content, viscosity and flow, are considered in this chapter. Methods of paint formulation are summarised and some aspects of colour technology are considered.

2.2 Pigment loading

2.2.1 INTRODUCTION

When discussing a paint formulation, the volume contributions of the pigments and extenders are used in preference to weight percentages because of the very wide range of densities of the various pigments, extenders and binders that are used in paints. In particular, a knowledge of the pigment volume concentration (P.V.C.) of the material enables a scientific approach to be adopted to the formulation of paints, and also to the subsequent interpretation of test data. When the performance of paints of dissimilar composition is being assessed, a knowledge of their P.V.C.s is essential for the accurate comparison and assessment of test results.

Although more precise information on a paint's performance is obtained by a consideration of the volume contribution of the pigments, there are instances where the more simplistic approach of expressing the weight relationships between the pigments and the non-volatile binder is acceptable. This relationship is known as the pigment-to-binder ratio. (In this chapter the term 'pigments' may be taken to comprise total pigment content including any added extender materials).

2.2.2 PIGMENT-TO-BINDER RATIO

A considerable advantage when considering the weight relationship of the pigment and binder content of a paint is the low level of mathematical manipulation required, compared to that involved for volume relationship, and this can be especially useful during the preliminary stages of formulation. The derivation of the pigment–binder ratio is detailed in Chapter 3.

The categorising of paint types by their pigment–binder ratios is possible in many instances and, furthermore, an indication of likely coating performance can be derived from this ratio. This is particularly useful when a paint of unknown performance has been analysed to obtain a percentage weight formulation for its basic ingredients, that is, total pigmentation, binder solids and solvent content.

A consideration of the example formulations given in Chapters 4–7 will provide an indication of the various pigment–binder ratios used for different applications. In general, finishes have a pigment–binder ratio of about

0·25–0·9:1·0, whereas primers can be grouped between about 2·0–4·0:1·0. Emulsion coatings of the type used for interior and exterior use on masonry can also be so classified, the exterior types typically being grouped in the range 2·0–4·0:1·0, and interior quality emulsion paints being in the region of 4·0–7·0:1·0. Many speciality coatings, however, cannot be readily categorised.

Formulations of high pigment–binder ratio are generally not suited for exterior exposure where extreme durability is required—a figure somewhere in the region of 4·0:1·0 probably represents the upper limit. This applies irrespective of the choice of pigmentation or binder type and is due to the inability of the relatively small amount of binder to form a continuous matrix around the large amount of particulate material present.

2.2.3 PIGMENT VOLUME CONCENTRATION (P.V.C.)

The information provided by a knowledge of the pigment–binder ratio is somewhat limited, especially when comparative performance tests are performed on paints of dissimilar composition. Knowledge of the P.V.C. of the formulation will facilitate a more scientific approach to the interpretation of test data since many physical properties of coatings respond in a well defined manner to compositional changes.

In practice, it is found that the addition levels of certain pigments into formulations can be classified into P.V.C. bands. Examples of the volume addition rates for some of the more common pigments are shown in Table 2.1.

The derivation of pigment volume concentration from the formula

$$\text{P.V.C. } (\%) = \frac{\text{volume of pigments and extenders}}{\text{volume of pigments and extenders} + \text{volume of solid binder}} \times \frac{100}{1}$$

is described in Chapter 3. The importance of P.V.C. can be demonstrated when paint is formulated to a range of P.V.C.s using similar constituents and preparative techniques, and is subjected to certain physical tests.

Critical pigment volume concentration (C.P.V.C.)

Many coating properties, especially those associated with porosity, such as permeability, blister and corrosion resistance as well as tensile strength and abrasion resistance, are found to exhibit sudden marked changes when a particular volume concentration of pigment in the film is exceeded (Figure 2.1). This formulation point is known as the critical pigment volume concentration (C.P.V.C.) and it is a fundamental characteristic of all pigmented coatings.

The C.P.V.C. of any coating represents the point at which, as the P.V.C. is increased, there is just sufficient binder to wet the pigment particles. At formulation points below C.P.V.C. there is a surfeit of binder so that the

Table 2.1 Typical Pigment Addition Levels

Pigment	P.V.C. (%)
White	
Titanium dioxide	15–20
Zinc oxide	15–20
Antimony oxide	15–20
White lead	15–20
Yellow	
Lead chromates	10–15
Zinc chromates	10–15
Cadmium sulphide	5–10
Hansa yellow	5–10
Benzidene yellow	5–10
Yellow iron oxides	10–15
Green	
Chromium oxide	10–15
Lead chrome green	10–15
Pigment green B	5–10
Blue	
Prussian blue	5–10
Ultramarine blue	10–15
Phthalocyanine blue	5–10
Red	
Red iron oxide	10–15
Toluidene red	10–15
Arylamide red	5–10
Black	
Black iron oxide	10–15
Carbon black	1–5
Protective	
Basic lead silicochromate	25–35
Basic lead sulphate	15–20
Calcium plumbate	30–40
Red lead	30–35
Zinc phosphate	25–30
Zinc tetroxychromate	20–25
Zinc chromate	30–40
Metallic	
Stainless steel	5–15
Aluminium	5–15
Zinc	60–70
Lead	40–50

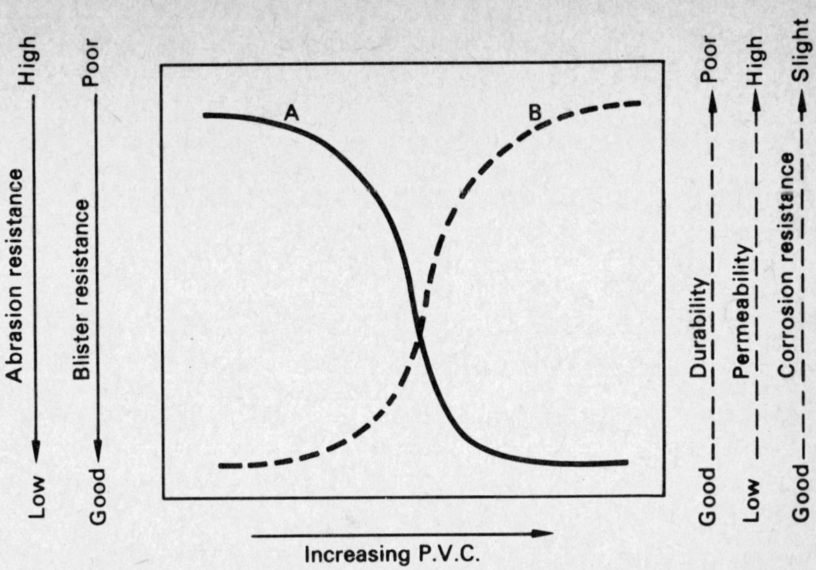

Figure 2.1 Influence of pigment volume concentration (P.V.C.) on paint performance

particles are firmly held in the matrix, while above the C.P.V.C. there is insufficient binder to wet the pigments which are therefore only loosely held in the film. For this reason it follows that high-performance or exterior coatings in general should not be formulated above their C.P.V.C. since many physical properties would be adversely affected. Conversely, in situations which are non-critical (generally interior), coatings formulated above their C.P.V.C. can be employed. An indication of the various P.V.C. levels adopted for certain paint types can be obtained from the formulations presented in Chapters 4–7.

The numerical value of the C.P.V.C. of a formulation varies according to the particular pigment and extender combination employed, and the important factors in this are the ability of the binder to wet the pigment and, in turn, the ease with which the pigments can be wetted by the binder. Generally, the more readily wetted types of pigment are conducive to low formulation C.P.V.C. Conversely, resins of high binding ability tend to increase the formulation C.P.V.C. Other factors are also important, however, such as the pigment particle size and shape as well as its distribution in the coating.

It is difficult to generalise on the C.P.V.C. of coatings since, as previously stated, this is determined by the exact formulation adopted. However, for many systems the C.P.V.C. lies in the region of 50–60 per cent P.V.C. The exact value of a formulation can only be accurately determined by empirical testing and by monitoring the change in the property being evaluated.

Table 2.2 Physical Data on Pigments and Extenders

Pigment	Specific gravity	Oil absorption (g/100 g pigment)
White		
Titanium dioxide	3·9–4·2	18–27
Zinc oxide	5·6–5·7	11–27
Antimony oxide	5·75	11–13
White lead	6·6–6·8	8–15
Yellow		
Lead chromates	5·8–6·4	12–25
Zinc chromates	3·4–3·5	24–27
Cadmium sulphide	4·2	25–35
Hansa yellow	1·4–1·5	40–50
Benzidene yellow	1·1–1·2	40–50
Yellow iron oxide	4·1–5·2	15–60
Green		
Chromium oxide	4·8–5·2	10–18
Lead chrome green	2·9–5·0	15–35
Pigment green B	1·47	60–70
Blue		
Prussian blue	1·85–1·97	44–58
Ultramarine blue	2·33	30–35
Phthalocyanine blue	1·5–1·64	35–45
Red		
Red iron oxide	4·1–5·2	15–60
Toluidene red	1·4	35–55
Arylamide red	1·4–1·7	40–60
Black		
Black iron oxide	4·7	20–28
Carbon black	1·7–2·2	100–200
Protective		
Basic lead silicochromate	4·1	13–15
Basic lead sulphate	6·4	10–14
Calcium plumbate	5·7	12–19
Red lead	8·9–9·0	5–12
Zinc phosphate	3·3	16–22
Zinc tetroxychromate	4·0	45–50
Zinc chromate	3·4	24–27
Metallic		
Stainless steel	~ 7·9	—
Aluminium	2·5–2·6	—
Zinc	7·06	—
Lead	11·1–11·4	—
Extenders		
Barytes	4·25–4·5	6–12
China clay	2·6	30–60
Mica	2·8–3·0	30–75
Talc	2·65–2·8	27–30
Whiting	2·53–2·71	13–22

This can clearly be a lengthy process and, of course, in many situations (especially when formulating at relatively low P.V.C.s) there is no real benefit in knowing the C.P.V.C. of the particular combination of ingredients used in the formulation. However, when formulating in the higher P.V.C. regions, it is important to know how far the system is from criticality. It is clearly undesirable that some relatively minor compositional change, say during the manufacturing process, could take the paint over its C.P.V.C.

Since the C.P.V.C. is a fundamental characteristic of any pigment then, similarly, pigment substitution on a volume basis can cause inadvertent changes in a formulation's C.P.V.C. over and above the intended compositional changes. As performance characteristics tend to change rapidly when the P.V.C. varies, particularly when it is close to the C.P.V.C., the modified performance patterns can be confused with the effects of the compositional changes.

Knowledge of a paint's P.V.C. in relation to its C.P.V.C. can therefore be a useful formulation aid. The value of this perhaps can best be realised by considering two paints having identical P.V.C.s, but which vary in their C.P.V.C.s in that one is under its C.P.V.C. (i.e. ratio of P.V.C. to C.P.V.C. $<1{\cdot}0:1{\cdot}0$), whereas the other is over its C.P.V.C. (i.e. ratio of P.V.C. to C.P.V.C. $>1{\cdot}0:1{\cdot}0$). Clearly, reference to Figure 2.1 shows that major performance differences would be expected, and hence the characteristics of a paint can be seen to be dependent not only upon its P.V.C. but also upon its C.P.V.C.

C.P.V.C. determination

Above the C.P.V.C., several physical properties of a paint film will change because it is underbound (there is insufficient binder to surround all of the pigment particles). The measurement of a property at different P.V.C.s will, provided that the test is carefully selected, reveal the C.P.V.C. of the formulation, and this empirical approach is often adopted.

With emulsion paint formulations, scrub- or stain-resistance as well as tensile strength measurements are suitable tests for identifying the system C.P.V.C. although, as with all such testing, these give a more or less narrow range within which the C.P.V.C. falls rather than a precise value. With solvent-soluble paints, similar empirical tests can be used. Such parameters have been used as the gloss level hold-out of a standard semi-gloss paint applied over the test paint at various P.V.C.s (the C.P.V.C. being indicated by a sudden drop in the measured gloss level of the topcoat), and changes in water vapour permeability.

All such empirical tests to determine C.P.V.C. are obviously very demanding of time and various more rapid techniques have been suggested. Many involve procedures over which there is no general agreement, but one such simple method is given here, and this has been shown to provide reasonably accurate information. The approach uses oil-absorption data for pigments, obtained by the standard spatula rub-out technique. Oil absorption is a measure of a pigment's wetting characteristics and this value is expressed as the quantity of linseed oil required to convert a given mass of pigment to a coherent, uniform paste. A pigment's oil absorption is largely dependent upon its particle size and shape.

Clearly inaccuracies are incurred both by using linseed oil, of which the wetting characteristics differ from those of many other types of binder, and in the poor reproducibility of the oil-absorption test, especially between different operators. However, this approach has the great merit of speed of testing. With this test, the critical pigment volume concentration can be calculated from the relation

$$\text{C.P.V.C. }(\%) = \frac{1}{1+\text{O.A.}} \times \frac{100}{1}$$

where the oil absorption (O.A.) is converted from the normal mode of expression, that is grammes of oil per 100 grammes of pigment, to millilitres of oil per millilitre of pigment. The oil absorption and specific gravity data for a range of pigments and extenders necessary to calculate the theoretical C.P.V.C. using this relation are shown in Table 2.2. However, the value derived is dependent upon the exact determinations for these test criteria obtained for a specific batch of raw material.

2.3 The solvent phase

2.3.1 INTRODUCTION

The third main component of a paint formulation, after the pigments and the binder, is the solvent. The solvent is the volatile constituent of a paint and its important functions include control of the viscosity of the liquid paint and hence of its application characteristics. Furthermore, as it evaporates from the applied film, the solvent (or blend of solvents) has an important role in controlling and modifying the drying speed and flow characteristics, as well as, with non-convertible coatings, the film-formation process which can have a major effect on the ultimate service performance.

In addition to these obviously major functions, solvents carry or solubilise certain of the resin components to facilitate paint manufacture. Furthermore, by judicious selection of the solvent or solvent blend, the formulator can control raw material costs as well as ensure that the formulation complies with any relevant legislation concerning environmental pollution, fire hazard or toxicity.

2.3.2 EVAPORATION RATE

The rate at which solvents evaporate from an applied film is an important factor in the drying, flow and film-formation process of a coating. It is of especial importance in formulations based on non-convertible polymer systems.

The rate of evaporation of individual solvents is often quoted as a numerical value relative to a standard solvent, which is normally either *n*-butyl acetate or diethyl ether. Certain mixtures of solvents, however, may have higher or lower rates of evaporation than their individual constituents as a result of azeotrope formation, and this may be experienced in many formulations where solvents of different classes are used.

Determination of evaporation rate

The simplest method of determining the evaporation rate of a solvent or mixture of solvents is to pipette a small volume onto a filter paper alongside an equal volume of a standard solvent and to record the time for complete evaporation of each. Alternatively, the rate of evaporation can be determined by periodic weighings of known volumes of solvent from a watch glass. By control of external factors such as air-speed, temperature and relative humidity, useful information on the likely performance of a solvent in a coating can be obtained. However, allowance must be made for the modifying effects on the evaporation rate of the other formulation components, especially the binder. The evaporation rates of some of the more commonly used solvents are given in Table 1.6 (page 27).

Effects of evaporation rate

During the evaporation of a solvent from an applied film, especially from non-convertible coatings, consideration has to be given to the effects on drying time. Too slow a rate produces unnecessarily long drying times while, conversely, too rapid an evaporation rate can result in poor flow-out. Several such problems due to evaporation rate are specific to spray-applied systems.

One of these problems is *solvent-blush*. This is caused by the rapid cooling effect as solvent evaporates from the applied film. Water vapour in the atmosphere in the immediate vicinity of the surface is also cooled and, if its vapour pressure is higher than the saturated vapour pressure of water at the surrounding ambient temperature, liquid water will condense onto the film itself. This water is then trapped in the coating, unable to escape as the paint viscosity increases during drying and, especially in clear films, the dissimilar refractive indices of the water and the film result in the coating acquiring a white discoloration, the so-called solvent-blush.

A further problem often experienced during spray application is *dry spray*. This results from the evaporation of a particular solvent from the atomised stream of paint before it can assist in the deposition and subsequent flow of the coating on the surface of the component being sprayed. The resultant film can be granular in appearance and this effect is the dry-spray phenomenon. Although this type of problem is often caused by inadequacies in the spray technique adopted, it can be minimised by correct solvent balance with respect to evaporation rate.

A problem specific to high-build coatings applied by airless spray, namely *vacuole (bubble) formation* within the film which leads to reduced service performance, has not yet been satisfactorily explained. There seems little doubt, however, that solvent balance and its effect on evaporation rate is the cause. The effect of solvents on this phenomenon is discussed in Chapter 4.

2.3.3 SOLVENT POWER

Some indication of the solvent power of a particular solvent (or combination of solvents) for a resin can be obtained by a number of laboratory methods, ranging from simple visual assessments of blends by checking for compatibility to viscosity measurements of solutions. With the latter approach, the most effective solvents for a given polymer, that is, those with greatest solvent power, are those which produce the lowest solution viscosities at constant solids contents.

With many formulations, the term 'solvent' needs some redefinition since specific partial or non-solvents for the polymer component may be added for reasons of performance or cost. Such polymers include vinyl resins (copolymers and dispersion systems), chlorinated rubber, polyurethanes and epoxies; many examples of this practice are given in Chapters 4–7. In such cases the terms 'primary (or true) solvent' and 'diluent' are perhaps more appropriate. The primary solvent has the property of dissolving the resin when used alone, but the diluent can be added to the solution of the resin in a primary solvent only in limited amounts if it is not to cause incompatibility and precipitation. The relative evaporation rates of such solvent–diluent combinations from applied films is clearly critical, and what must be avoided is the situation where the content of diluent in the drying film increases to a point where the binder precipitates, since this can clearly result in reduced performance.

The precipitation point can be determined experimentally by adding the diluent to a known amount of the binder dissolved in the requisite primary solvent. During formulation using combinations of primary solvent and diluent, it is unwise to use a mixture of the two that is close to the precipitation point.

2.3.4 FLASH-POINT

The flash-point of a flammable liquid is the lowest temperature at which its vapour can be ignited in the presence of a flame or spark. The flash-point of a liquid is related to its vapour pressure, that is, the boiling-point, although compositional details exert a marked influence in the case of paints. Although there is no theoretical basis upon which flash-points of solvent mixtures can be calculated precisely, a good rule of thumb is that the flash-point of a mixture is close to that of the component with the lowest flash-point, provided it constitutes a major proportion of the mixture.

The flash-point of a paint or solvent is therefore an indication of its flammability, and there are many national and international regulations controlling transport and storage of flammable liquids. The measurement and control of flash-point can therefore be an important requirement of a coating formulation, and there are several standard methods of determining this characteristic.

The widely used closed-cup technique involves introducing a small flame into the atmosphere above a volume of the test paint the temperature of which is being slowly raised. The whole test is necessarily performed under carefully controlled conditions, and the flash-point is taken as the lowest temperature at which the solvent vapour and air mixture ignites (flashes) on the introduction of the flame. The flash-points of some of the more commonly used solvents are given in Table 1.6 (page 27).

2.4 Viscosity

2.4.1 INTRODUCTION

The viscosity of a complex system such as paint is governed by the interaction of all of the ingredients in the formulation. Of major importance, however, are the solvent composition, and its content, as well as the presence of viscosity control agents, commonly called thickeners or gellants. The viscosity of a paint formulation controls the storage stability, the application characteristics and the subsequent flow that occurs immediately after application. Application and flow characteristics need to be optimal for the particular system if the coating is to perform satisfactorily during service.

2.4.2 APPLICATION AND STORAGE CONSIDERATIONS

The viscosity adopted for a paint at the formulation stage is determined largely by the storage requirements as well as its application characteristics. For example, industrial users of bulk paints frequently purchase, and store, a viscous product which subsequently can be diluted to its application viscosity immediately prior to use with an appropriate solvent. Conversely, in many other applications, paints are preferred ready for use straight from the container. Generally, ready-for-use formulations are for site application and for the large retail markets. One reason for this difference is that coatings supplied in a viscous state and diluted before application, e.g. by spray, would generally not store well at their final viscosity. This is mainly because pigment settlement may be excessive, resulting in poor colour uniformity as well as increased labour costs in reincorporating the settled material satisfactorily. Paints supplied ready for use, however, are typically applied using those techniques where more viscous products are required, that is, by brush, roller and airless-spray methods, and hence storage stability is easier to achieve.

The viscosity of a paint therefore has to be matched to its intended application technique as well as having a value that ensures that it can be stored in a stable, usable condition after manufacture and prior to use. It is difficult to categorise different paints into viscosity groupings, but many solvent-soluble coating types are formulated somewhere in the region of 4–6 poise (0·4–0·6 N s m^{-2}). This viscosity permits brush, roller and airless or hot spray application without further addition of solvents. However, for conventional spray application, dilution with a suitable solvent to about 1 poise (0·1 N s m^{-2}) would be necessary. Many of the formulations shown in Chapter 4, especially those industrially applied, would require dilution if application was to be by this type of spray equipment.

Thixotropic behaviour

Many coating types are formulated in a thixotropic condition since this has several practical advantages. Thixotropic paints have a gelled structure, the rigidity of which can be varied by selection of the thixotrope used. A detailed discussion of paint rheology is outside the scope of this text, but a few comments may be useful here. Thixotropic paint systems have what can be considered to be a time-dependent viscosity behaviour. In this respect they differ from paint, of which the viscosity is constant irrespective of the time taken to perform the measurement.

The simplest type of flow property is Newtonian flow: the viscosity of Newtonian systems is invariant with respect to both time and the shearing rate of the liquid (during viscosity measurement, stirring or application), although the viscosity is temperature-dependent. Newtonian liquids are essentially simple systems such as water, solvents, mineral oils and a few low-molecular-weight resin solutions.

Most paints, however, are non-Newtonian systems. The types of non-Newtonian flow observed by the paint technologist include dilatant flow, plastic flow and pseudo-plastic flow. Thixotropic systems can be considered to be associated with plastic and pseudo-plastic flow.

With dilatant flow, the apparent viscosity increases with increasing rates of shear, clearly an undesirable property with paints. Such rheological behaviour is generally associated with systems of very high solids content such as fillers and stoppers, although dispersion coatings (organosols and plastisols) can also exhibit this behaviour. The layer of pigment settlement obtained after prolonged storage of liquid paints is also frequently dilatant in nature.

Plastic flow is said to be exhibited when there is a certain minimum stress, known as the yield stress, which must be overcome before flow can take place and, once this has been exceeded, the viscosity behaviour is essentially Newtonian. Such systems are sometimes referred to as Bingham bodies, after one of the researchers who first identified this behaviour. Pigmented systems can show plastic flow although the binder itself must have Newtonian flow characteristics.

Systems with pseudo-plastic flow have no yield value—if represented graphically, the curve of shear stress against shear rate is non-linear, with the shear rate increasing faster than the shear stress. Pseudo-plastic flow has been interpreted as being due to the alignment of long-chain molecules by an applied shear stress, and is exhibited by certain emulsion paint systems.

The viscosity of materials showing these types of flow behaviour (dilatant, plastic and pseudo-plastic) remains substantially the same irrespective of the time for which shearing is applied. In contrast, the viscosity of thixotropic materials depends on the duration as well as on the rate of shear. Thixotropy is a reversible process, that is, after removal of the shear stress the viscosity gradually builds up again. Many paint systems exhibit a degree of thixotropy which is usually obtained by the inclusion of specific agents, the action of which will be considered later. The formation of a thixotropic body has been explained as being due to the increasing orientation of particles and molecules during periods of rest, which produces a viscous gel state. When shearing occurs, however, the linkages effecting the orientation are broken and the viscosity falls until, at equilibrium, constant viscosity is obtained. At this point, the processes of linkage cleavage and reformation are occurring at equal rates. When the shear stress is removed, that is, after stirring and, more importantly, after application of the paint, the viscosity returns to its initial value. This occurs at a rate which is dependent on the strength of the secondary valencies causing the linking reactions within the system, and the ease with which the molecules can re-orientate themselves.

Clearly thixotropy can impart useful attributes to a paint formulation. Settlement of pigments can be entirely eliminated and this also reduces the need to stir paints prior to application. Since flow under low shear conditions is reduced (as thixotropy develops after application of the paint), somewhat greater wet film thicknesses can be deposited and, furthermore, running and sagging from vertical surfaces is reduced.

2.4.3 VISCOSITY CONTROL

Many of the methods of viscosity control involve the addition of thickeners to paint formulations to introduce varying degrees of thixotropy into the system. Exceptions to this are the cellulosic thickeners added to emulsion paints; these do not induce true thixotropic characteristics but rather impart more pseudo-plastic behaviour. In systems where thixotropy is not required, for example in high-gloss finishing paints where flow needs to be maximised (thixotropic systems rarely have the same ability to flow as non-thixotropic paints, and hence the surface finish tends to be somewhat inferior), the viscosity is controlled to the desired limits by selection of the binder and by adjusting the total solids content of the formulation.

The main method of inducing thixotropic characteristics into solvent-borne formulations is by the inclusion of gelling agents such as the amine-treated clays and hydrogenated castor oil; the mode of incorporation of these materials is discussed later. Alkyd binders gelled by the incorporation of specific polyamides during manufacture are also used in finishing systems, especially for the retail market.

Various addition levels of the gelled alkyd can be used in conjunction with a non-gelled, conventional oil-modified alkyd binder in order to control the thixotropic nature of the resultant paint formulation, and levels of 10–100 per cent of the total binder can be used. With the clay types of additive, however, levels are much lower and rarely need to exceed 2–3 per cent of the total formulation weight. Adjustment of the structure of the paint with these materials is controlled by the addition level.

With emulsion systems, much higher viscosities are generally adopted than those for solvent-soluble systems, with viscosity levels commonly in the region of 10–15 poise (1–1·5 $N\ s\ m^{-2}$). Although the cellulose ether thickeners most frequently used do not induce thixotropy as such, the highly structured nature of these formulations provides many of the advantages of a thixotropic system. Flow-out, however, is not especially good with emulsion systems thickened with cellulosic types of thickener, although the use of the ionic types, such as sodium carboxy methyl cellulose, does result in better flow properties in the applied film compared with non-ionic grades such as ethyl hydroxy ethyl cellulose.

With these cellulose materials, paint viscosity is controlled not only by the addition level in the formulation but also by the solution viscosity (molecular weight) of the grade adopted. Typically, the cellulose thickener is pre-gelled by vigorous stirring into water at a concentration of 2–3 per cent. A quantity of this pre-gel is then added into the formulation, generally at the pigment-dispersing stage of manufacture, at a level sufficient to impart the required viscosity. Depending on the grade, addition levels are typically up to 1 per cent of cellulose ether on total formulation weight.

2.4.4 VISCOSITY MEASUREMENT

Many different techniques are used to measure the viscosity of paints and, clearly, the more complex rheological systems require sophisticated approaches if precise information on shear rate–shear stress relationships is required. However, for the more routine work on formulation, quality control, etc., the two most widely used techniques for measuring the viscosity of a formulation are flow (efflux) cups and rotational viscometers.

Flow cups are used for relatively low-viscosity materials, although strictly they should only be used for Newtonian systems. They are widely employed in production control, during both paint manufacture and paint application. These viscometers are available with various sizes of orifice so that a range of viscosities can be measured. The viscosity of a particular paint is quoted as the time (in seconds) for the liquid to evacuate the cup, falling under gravity, through the orifice in the base.

Higher viscosities can be measured more accurately with rotational viscometers than by flow cups and the read-out is generally directly calibrated in appropriate units of viscosity, for example, the poise. The underlying principle involves moving a paddle or other shearing device through the paint and measuring the resultant shear stress generated. Many of these viscometers are highly sophisticated and can be used for research applications (and are essential for characterising thixotropic systems), although simple constant-shear-rate devices are widely used for general purpose work.

The most widely used unit of viscosity is still the poise (having the symbol P) or its subunit, the centipoise (cP). In the SI system the unit of viscosity is the Newton second per square metre: $1\ \mathrm{N\ s\ m^{-2}} = 10\ \mathrm{P}$. These values are known as dynamic units of viscosity. Occasionally the so-called 'kinematic viscosity' is encountered, which is obtained by dividing the dynamic viscosity by the density of the liquid; the unit is the stoke (St) or centistoke (cSt). This particular unit, however, is less used in modern practice. The relationship between commonly used units of viscosity is shown in Figure 2.2.

2.5 Methods of formulation

2.5.1 INTRODUCTION

A consideration of the subjects outlined in the preceding sections of this chapter gives an indication of the various factors that need to be taken into account in the formulation of a paint. Working to the various guidelines presented both here and elsewhere in this text enables a systematic approach to be adopted to the formulation of many coating types. For example, selection of the polymer type and pigmentation is clearly dependent upon

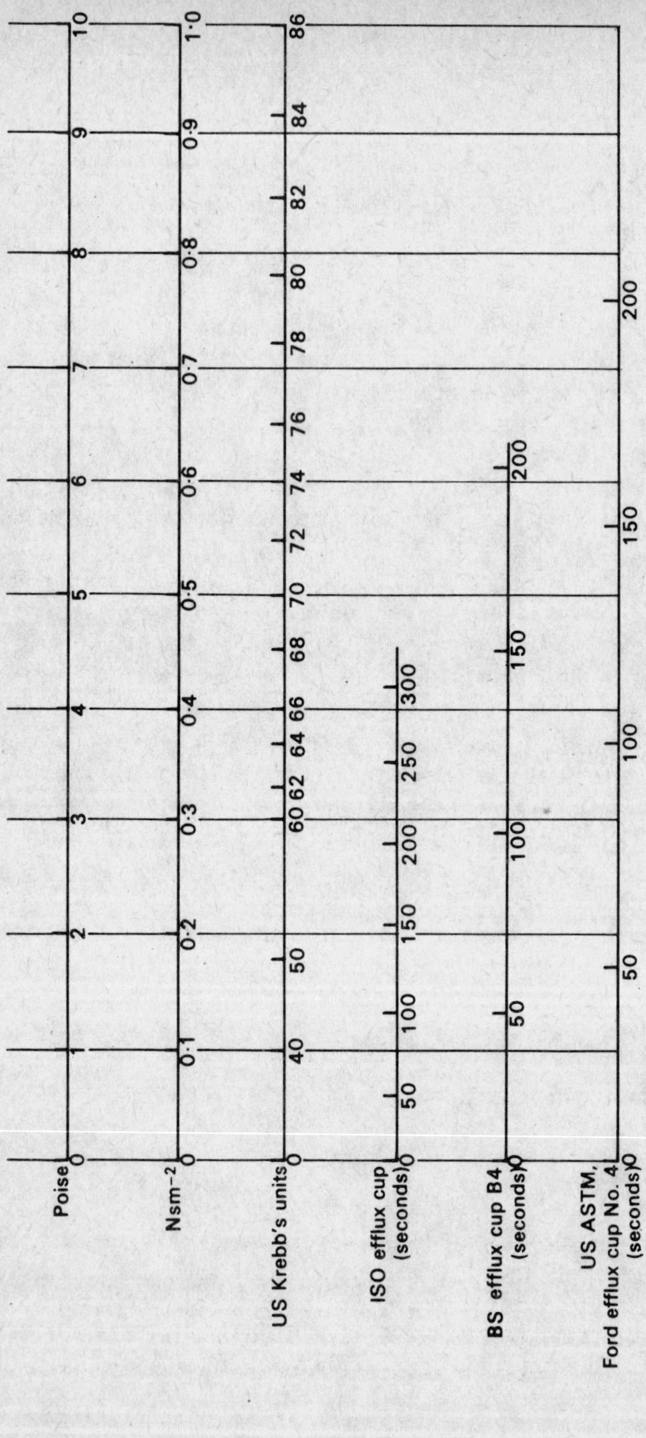

Figure 2.2 Viscosity unit conversion chart

the anticipated service requirements, and obviously the nature of these major components will govern the selection and use of many of the ancillary additives such as driers and viscosity modifiers. Addition levels of pigments are clearly governed by certain well defined principles although, obviously, novel coatings can be formulated outside these guidelines.

While the formulation of a paint is relatively easy, to optimise the composition to enhance a specific property, or combination of properties, within the rather broad limits within which most paint types can be formulated is rather more complicated. It is in this area that certain techniques for the manipulation of formulations can be usefully applied in order to ensure that a coating exhibits all of the attributes required of it.

2.5.2 STATISTICAL METHODS

A statistical approach to the design of experiments to optimise a paint formulation has certain advantages. Perhaps most importantly, random errors, which can be large with the testing of complex multi-component systems such as paints, can be reduced to an acceptably low level. Furthermore, the effects of a number of variables on performance can be precisely evaluated. There are drawbacks, however, not least of which is the high degree of mathematical ability needed to set up and analyse the series of experiments correctly. Another problem with the statistical approach is the large number of individual tests which may be required. In a complete factorial experiment, for example, all combinations of all levels of the formulation variables have to be tested and if repeats are necessary then, in order to maintain the balance of the experiment, all must be repeated.

Nevertheless, in certain situations a statistical approach to formulation optimisation is necessary, although frequently considerations of time require that such experiments are reduced in size or scope. There are, however, statistical processes that ensure that too much information is not lost by the reduction in experiment size.

Statistical methods are also of particular value in the analysis of test data. The techniques that can be used for this purpose are discussed in Chapter 3.

2.5.3 NON-STATISTICAL METHODS

In many respects, non-statistical approaches to paint formulation optimisation have much to commend them, particularly speed. However, there is some loss of information and precision and, clearly, these two considerations have to be balanced before deciding on a particular approach to formulation testing.

One of the simplest methods of optimising a paint formulation involves studying the effect on paint performance of changes in two composition variables. With this technique, two variables (that can be used at different

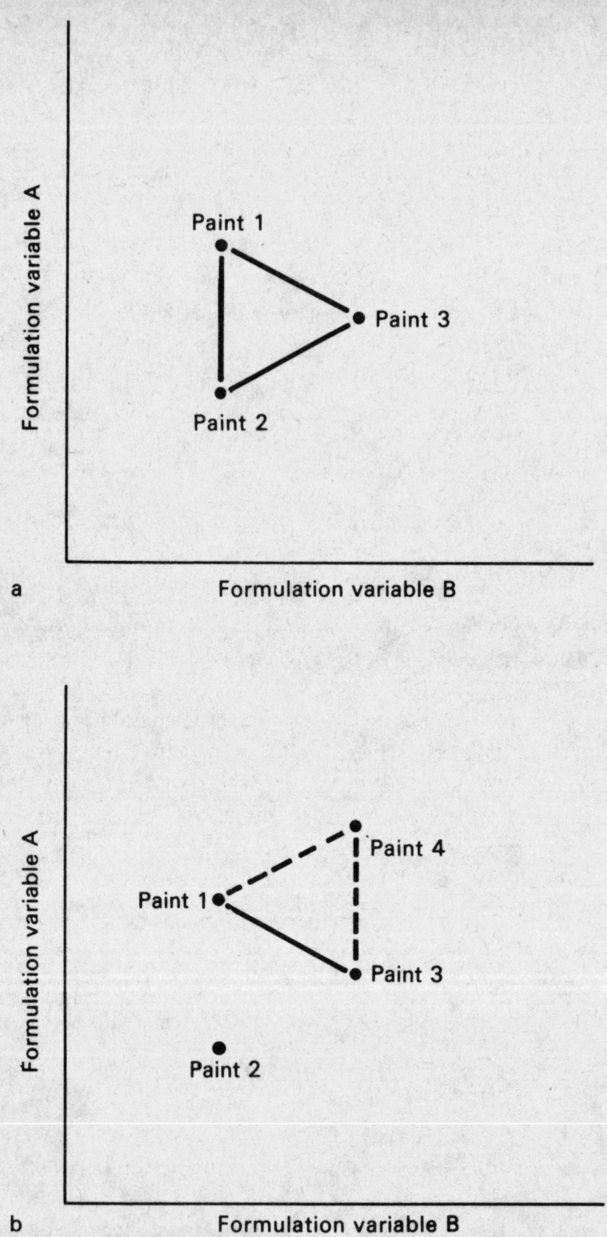

Figure 2.3 Graphical technique of paint formulation
a Initial formulation points
b Formulation point 4 is plotted after Paint 2 has been rejected

addition levels) are selected—clearly the right ones must be chosen to avoid time wastage. Other formulation factors are then kept constant throughout the test; thus one variable may be plotted along the abscissa of a graph, and the other along the ordinate. Typically, the pigment volume concentration and resin binder content might be selected as the variables. An equilateral triangle is then constructed on the graph and paint formulations are prepared corresponding to the coordinates of the three corners of the triangle (Figure 2.3a).

Tests are undertaken on the three paints, and the formulation with the worst performance is rejected. A fresh triangle is then constructed positioning the new co-ordinate at the mirror image position of the rejected formulation point (Figure 2.3b). Tests are again performed on the paints and the above procedure is repeated until the area of optimum performance is reached. This is indicated by the circling motion of the triangle around a certain area.

Although this is a very simple test procedure, some skill is required not only in selecting the right variables, but also in determining the scale length along the axes as well as the length of the arms of the resultant triangle. Assessment of the results must also take into account the fact that, with paints, improved performance in one area is often gained at the expense of other properties. This point, however, is also a consideration even with fully statistical approaches to paint formulation.

This very simple two-variable test can be extended to the study of four or more variables although the mathematics involved tends to become difficult.

2.6 Pigment dispersion

2.6.1 INTRODUCTION

Having derived an optimum formulation, it is necessary to manufacture the paint. This basically involves effectively dispersing the pigments and extenders into the binder phase.

The dispersion process is important in several respects since many coating properties are dependent upon the degree of dispersion of the pigmentation. Typically, it affects storage properties as well as attributes that are important during service such as colour and gloss, and even the durability of the paint film. Dispersion is also important during manufacture since the ease with which a system can be prepared must be matched to production targets. A consistent degree of pigment dispersion is also clearly beneficial in maintaining the reproducibility of colours during a manufacturing programme. Furthermore, the reagglomeration of pigments in a finished paint during storage can result in colour drift as well as problems of flotation (surface mottling of the dried film caused by segregation of the pigment components during drying).

2.6.2 MECHANISM OF DISPERSION

The process of dispersing a pigment in its binder consists of a number of well defined stages, although obviously these occur concurrently during manufacture.

First, and most importantly in that this stage requires the greatest expenditure of effort, is the wetting of the pigment surface by the binder. This involves displacing adsorbed gases and water from the surfaces of the pigment and extender particles and replacing them with the binder phase. Secondly, the agglomerates of pigment particles have to be broken down into their individual particles if the pigment is to be utilised to its fullest effect. The final stage involves keeping the separated particles in a stable condition so that they do not reagglomerate (flocculate) during storage.

Stabilisation of a system against pigment flocculation is achieved either by the presence of electrical charges on the pigment particle surfaces, which leads to mutual repulsion, or by the presence of adsorbed layers, for example of polymer, which prevents close packing of the particles. The latter mode of stabilisation occurs largely in solvent-borne paint systems, whereas both mechanisms can occur in aqueous-based paints.

2.6.3 SURFACE-ACTIVE AGENTS

The wetting and stabilisation of pigments and extenders in paint formulations can be facilitated by the incorporation of surface-active agents, also known as surfactants or wetting agents. These have the characteristic property of being strongly adsorbed onto pigment surfaces, thus modifying the ease with which the dispersion process can be performed. Surface-active agents are generally only required at low addition levels (about 1 per cent or less of the total pigmentation) and they tend to be highly specific in their efficacy.

Four main types of surface-active agent are available, these being classified according to their ionisation characteristics. *Anionic* surfactants ionise in solution and impart an overall negative charge to surfaces; examples are the fatty acid sulphonates and salts of the higher fatty acids. *Cationic* surface-active agents also ionise but with them an overall positive charge is imparted to surfaces. Examples of cationic surfactants are the quaternary ammonium compounds. Although their main use is in the medical field by virtue of their bactericidal action, they are sometimes used as water-displacing additives in solvent-soluble primers.

Non-ionic surface-active agents do not ionise in solution and are most often based on ethylene oxide derivatives such as condensates of an alkyl phenol and ethylene oxide. Varying the chain length of the condensate yields materials suited for use with various binder types. The last group of surface-active agents are the *amphoteric* materials. These are molecules containing free amino and carboxyl groups and they ionise in solution to impart a net positive or negative charge depending on the solution pH. An example of this type of surface-active agent is soya bean lecithin.

The selection of a suitable surface-active agent for a particular paint formulation is mainly empirical owing to the high specificity of these materials. However, the charge imparted to the pigment surface must match that of the binder as this will reinforce the stability of the system. Should the charges on the polymer and pigment be opposite in sign, then charge neutralisation might lead to instability and precipitation.

Dispersion-promoting properties

In many instances, however, the use of specific surface-active agents to facilitate pigment dispersion is not necessary. In particular, the dispersion of most pigment types into certain binders, especially oil-modified alkyds, is relatively easy because the polymer contains components which promote the wetting process. In the case of alkyds, it is the presence of the highly polar carboxyl and, to a lesser extent, hydroxyl groups which facilitates the wetting process. The presence of these groups accounts for the acid value of the binder: alkyds of higher acid value have a greater pigment-wetting ability than those of lower acid value. Several other types of binder exhibit good pigment-dispersing ability, the extent depending on the polar nature of the binder's molecular structure.

Furthermore, many pigments are manufactured with modified surface properties so that their subsequent dispersion into paints is facilitated. With some treated materials only low energy techniques, in some instances simple stirring, may be required to effect dispersion.

An example of a pigment which is available with a wide range of surface treatments is titanium dioxide. If the grade selected for a particular application is matched carefully to that of the binder, dispersion should be readily accomplished without the need of any, or at least only very small amounts of, further additives.

With titanium dioxide, both inorganic and organic treatments are practised by pigment manufacturers, and both improve dispersibility into specific polymers as well as imparting other beneficial properties. In particular, the inorganic surface treatment of titanium dioxide (especially the rutile grade) with the oxides and hydroxides of aluminium, silicon, titanium or zirconium improves the durability of the resultant formulation. This is because they reduce the catalytic activity of the pigment, that is, the treatment reduces chalking (erosion) tendencies. Untreated titanium dioxide pigments are also available and these are characterised by rather lower durabilities as well as inferior dispersion characteristics. They are, however, of lower cost and find applications in interior paints.

A method of reducing the tendency of pigments to flocculate after dispersion is through the use of high paint viscosities. This reduces the ability of the particles to reagglomerate and, although this is not practical with many formulations, the technique works well with emulsion and certain solvent-borne systems.

2.6.4 PAINT MANUFACTURE

The two main classes of paint considered in this text, solvent-borne and aqueous (emulsion) paints, differ in the manner by which pigment dispersion is achieved.

Basically, solvent-borne paints can be prepared using a far wider range of equipment than emulsion paints, which tend to be made using high-speed impeller machines. With solvent-borne systems, ball mills are the traditional, and still relatively widely used, method of preparation, although increasingly sand or bead mills are coming into use. With solvent-borne coatings the composition of the mill base (that part of the formulation which is subjected to the initial dispersion process) is somewhat more critical than with emulsion coatings. Generally, optimum dispersion is only obtained if a predetermined ratio of pigment, binder and solvent is used.

The technique of determining this combination of ingredients to optimise the mill base composition is known as the Daniel flow-point method, and it has been successfully adapted for most combinations of formulations and manufacturing equipment. In essence, the flow-point technique is a modification of the oil-absorption test for a pigment or extender (pages 42, 73). A pre-weighed amount of the pigment in the formulation is successively titrated with different solution concentrations of the binder in its appropriate solvent. The paste formed from the two is stirred with a glass rod, and the point at which the paste 'snaps back' when the rod is pulled from the mass is noted with respect to the volume of resin solution added to the pigment, and the percentage resin concentration in solution. A graphical plot of these two variables shows, for a range of resin concentrations, a minimum volume requirement at one particular resin concentration (Figure 2.4). Suitably scaled up, that particular volume and addition serve as the basis for the mill base composition.

After the initial dispersion process has been performed, the pigment paste is 'let down' with a quantity of binder to stabilise the dispersion. This involves adding a proportion of the binder concentrate to the mill base as soon as possible after dispersion in order to ensure that flocculation (reagglomeration) does not occur. Finally, the remaining binder and solvent, and other minor additives, are added.

In emulsion paint manufacture using high-speed impeller equipment, the pigments are sifted into a quantity of water (to which the dispersing agents have been added) during shearing. The thickener can be added either before or after dispersion. The polymer component is blended-in under low shear

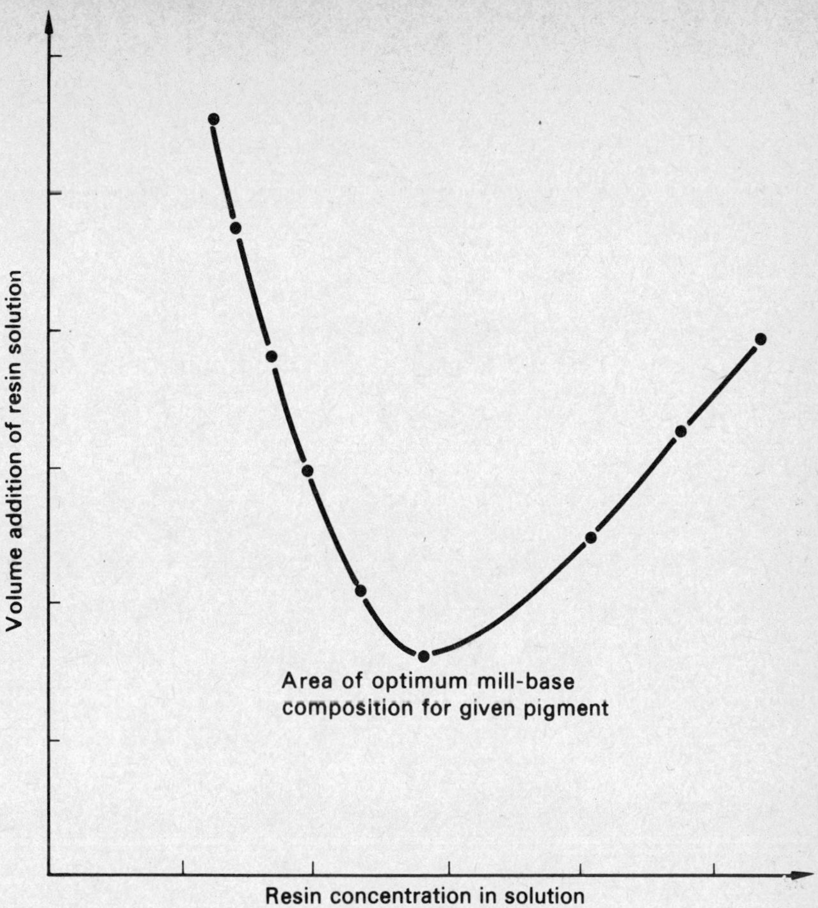

Figure 2.4 Typical Daniel flow-point curve

rates only after dispersion has been achieved. The reason for this is that high shear rates can destabilise many emulsion polymer systems. Other additions of water, coalescing solvents, etc., are made after the emulsion polymer has been incorporated, again with slow-speed stirring.

The consistency of the mill base has to be matched to the particular type of equipment used and this can be readily accomplished by changing the amount of thickener added at the dispersion stage, or by altering the solids content. However, pigment loadings are normally kept as high as possible while consistent with the mill base being mobile in the machine. A problem with emulsion paint mill bases is that they tend to be 'falsebodied'; this can result in poor circulation round the impeller mixing zone, and hence there can be poor dispersion. With large-scale production equipment this problem can be minimised by the use of tall, narrow mixing vessels with internally mounted flow-inducing vanes.

2.6.5 ASSESSMENT OF DISPERSION

There are many methods for assessing the degree of dispersion of paint mill bases. Many of these involve sophisticated instrumentation providing information on particle size and distribution.

Within most manufacturing facilities, however, the Hegman gauge is used, although this simple piece of equipment only provides data on the maximum particle size within the dispersed mill base. With the Hegman gauge, a small quantity of the paint mill base is drawn down a tapered, calibrated channel (typically graduated from 100 μm to 10 or 0 μm) and is then viewed obliquely. The dispersed particles are expressed from the liquid paint and, by reference to the scale, the channel depth at which particles are first evident gives an indication of the degree of dispersion. Typically, a primer coat or a low gloss paint such as an undercoat or eggshell finish paint, as well as most emulsion paints, has a maximum particle size of about 50 μm. In contrast, a full gloss decorative finish would need to be 'off the scale', that is, no particles should be evident.

Hegman gauges are widely used for solvent-borne systems, although they are less satisfactory with emulsion paint mill bases, unless these are 'let down' prior to test with a proportion of the emulsion binder component.

2.7 Colour technology

2.7.1 INTRODUCTION

In many instances paints need to be formulated to satisfy a colour specification, and this frequently involves the use of combinations of colouring pigments.

When colouring pigments are used in admixture, factors such as tint strength and mass colour, metamerism as well as lightfastness and the possibility of adverse chemical interactions must be considered at the formulation stage.

2.7.2 COLOUR MATCHING

Colour matching involves the adjustment of a paint's composition so that it conforms to an agreed colour standard. In practice, this involves formulating a paint so that after the pigment dispersion stage the colour obtained closely approximates to the standard. The colour is then adjusted by careful stepwise addition of predispersed pigment 'tinters' to the paint until the colour is judged an acceptable match. This final matching can be gauged either by the trained eye or by various instrumental techniques.

It is essential that visual colour matching is performed under standardised illumination, ideally in a purpose-designed colour-matching booth or cabinet. These cabinets are normally equipped with different types of standard illuminant so that metameric colour matches can be detected. Metamerism is the phenomenon whereby the apparent colour of a surface (typically one carrying a paint film) can vary depending on the nature of the light source. Thus a colour may be a close match to a standard under, for example, daylight but not under artificial light. This effect often arises when the combination of pigments used to prepare the paint differs from that used to prepare the colour standard. Metamerically different paints have different spectral reflectances; it is only when colours have identical spectral characteristics that they will match under all types of illumination. This necessitates that a coloured paint should always be prepared from the same combination of pigments.

Before pigment combinations for a particular colour can be derived, however, other factors related to the subsequent performance of the paint must also be determined. Certain characteristics can clearly be inferred from a knowledge of the properties and chemical structure of the pigment, although others need to be determined experimentally. Among the latter are the lightfastness of the pigment combination adopted and the possibility of any adverse interactions.

2.7.3 MASS COLOUR AND TINT STRENGTH

Both mass colour and tint strength of pigments can be regarded as properties which provide data on colour consistency. Testing of both properties would normally be performed as part of a quality control process on individual pigments prior to use but, in a slightly modified form, such tests can also be used to provide an indication of the amount of pigment required to produce a given shade.

The mass colour of a pigment is obtained by preparing a paste with an appropriate binder and comparing this with an agreed standard of that pigment. For this determination, it is important that identical ratios of binder to pigment and also comparable dispersing techniques are used. Dispersion can be effected by hand, using a spatula to rub out the paste on a glass plate, or with an automatic muller.

The tinting strength of a coloured pigment is a measure of its ability to colour a standard paint or pigment. The measurement technique is similar to that of determining mass colour, but with a known amount of the colouring pigment being added to a known weight of a white pigment or paint. Knowledge of any variation in the tint strength of a pigment can be used to adjust formulations prior to manufacture, and this is obviously an important factor in reducing batch rejection rates.

With slight modification, the tint-strength test can be used to provide guidance on the ratios of pigments required to match a given colour standard. The weights used can then be scaled by an appropriate factor for inclusion into the formulation mill base. Colour development is always greater during production than is obtained during dispersion by hand or muller, and some reduction in the amounts of pigments added is generally justified, especially where organic pigments of high tint-strength are used.

2.7.4 LIGHTFASTNESS

A property requiring experimental measurement and determination is the colour stability of paints under ultra-violet light, or lightfastness. Since various factors associated with the nature of the pigments, their relative ratios and the nature of the polymer (or binder) component can modify lightfastness, it is not possible to give a stability rating for pigments. As a general rule, however, the lightfastness of a pigment is decreased on increasing dilution with other pigments, especially when used in conjunction with white pigment.

The testing of colours for lightfastness must be performed using the paint formulation in which they will be used. Lightfastness testing is performed in cabinets containing a carbon arc light source. The colour stability of the particular pigment or combination is rated against unexposed samples.

2.7.5 STABILITY

Various phenomena can affect the colour stability of paints during storage of the liquid paint, upon application or subsequently during service. Many of these need to be determined experimentally at the formulation stage.

The compatibility of pigments with other components of the formulation may sometimes be limited, and this can result in colour changes during storage of the paint. Typical of these problems are the adverse interactions between sulphide-containing pigments and mercury- or lead-containing compounds, which result in a black discoloration, as well as the colour discharge exhibited by Prussian blue pigment when used in conjunction with binders containing unsaturated acids (in this latter combination, however, the effect is temporary since the full colour is restored on exposure to air).

A colour change frequently encountered during application arises from flotation or the associated phenomenon of flooding. Either effect may be found when the coloured pigments in a paint differ markedly in their specific gravities. During drying, convection currents induced in the film by solvent evaporation tend to bring the lower-density pigments to the surface. This effect is known as flooding and the surface layer takes on the general colour of the affected pigment; it develops in situations of low air flow. In contrast, the mottled surface colour characteristic of flotation is more likely to occur where air is moving over the film.

Both of these effects are characteristic of flocculated systems and the problem can be minimised in several ways, typically by varying the pigments selected and/or by the use of specific surface-active agents. These phenomena are also less likely to occur in more viscous formulations.

Several effects can cause colour instability during exposure of the paint, the most common being changes induced by light (see above) and those induced by chemical interactions. Information on many of the problems of chemical interaction can be obtained from the literature or by experiment, and this knowledge should be applied at the formulation stage. A typical example is the reduction of lead chromes by atmospheric sulphur dioxide, which can result in yellows turning white, and chrome greens turning blue.

Colour changes in pastel-shade paints containing titanium dioxide pigment can be attributed to chalking. The phenomenon is caused by a photochemically induced interaction between the white pigment and its binder, resulting in progressive breakdown of the latter. This leads to the progressive release of small amounts of the white pigment onto the surface of the film, and so causes a lightening of the colour. Chalking can be heavy with the anatase form of titanium dioxide and, for this reason, its use is largely restricted to paints for interior use. The rutile form forms chalk only at a comparatively slow rate and its use is therefore favoured in exterior situations.

Further details of specific interactions which limit or restrict the use of certain pigments, as well as other paint components, are given where appropriate in Chapters 4–7.

2.8 Bibliography

Morgan, W.M., *Outlines of Paint Technology*. George Griffin: High Wycombe, 1969.

Nylen, P. and Sunderland, E., *Modern Surface Coatings*. Interscience: London, 1965.

Gardner, H.A., *Physical and Chemical Examination of Paints, Varnishes, Lacquers and Colours,* Special Technical Publication 500. ASTM: Philadelphia, 1968.

Newton, D.S. (Ed.), *Paint Technology Manual,* Part 5, *The Testing of Paints*. Chapman and Hall: London, 1965.

CHAPTER 3

Mathematics in Paint Formulation

Paint technology involves the practitioner in a variety of specialised computations. Many of these are straightforward: others are more complex.

3.1 Paint formulae

An area of confusion in paint technology is the method of presenting the formulation. In paint works and production departments, it is common to use volumes for fluids such as liquid resins and solvents, and weights for solid additives. Percentage weights, however, are often quoted in textbooks and journals, while percentage volumes may be used when considering formulations for particular service conditions or in the calculation of dry film thicknesses.

In this book weight percentages have been used to ensure uniformity. However, it is useful to discuss here the conversion of one system of presentation to another, particularly with regard to problems of works production.

An initial example might be the batch sheet for the preparation of 250 litres of white interior–exterior emulsion paint. In practice, a slight excess is normally produced to allow for manufacturing losses so that the actual production volume will be 260 l.

Example A: Production calculation–quantity and unit cost

The paint formulation for this example is given in Table 3.1. The first step is to calculate the volume contribution of each component by dividing the percentage weight by the component density. At this stage, the theoretical density of the paint can be obtained by dividing the total weight of the paint by its volume:

$$100\ \text{kg} \div 70{\cdot}71\ \text{l} = 1{\cdot}414\ \text{kg/l}$$

Table 3.1 Percentage Weights and Volumes of a White Emulsion Paint

	Weight (%)	Density (g/cm³)	Volume (cm³)
Rutile titanium dioxide	19·2	4·20	4·57
Whiting	21·7	2·80	7·75
Emulsion resin	27·9	1·02	27·35
Thickener	0·7	1·33	0·53
Dispersing agent	0·5	1·0*	0·50
Biocide	0·3	1·0*	0·30
Coalescing solvent	0·5	0·98	0·51
Water	29·2	1·00	29·2
	100·0		70·71

*Assumed values.

Since the weight of 1 litre of paint is 1·414 kg, a batch of 260 l will weigh 260×1·414=367·64 kg, which can be rounded off to 368 kg. This total weight can be used to calculate the weight contribution of each component.

The percentage weight formula is converted to a total of 368 kg and the contribution of each component is obtained by multiplying its weight percentage (y) by 368/100, that is $y \times (368/100)$, as shown in Table 3.2. This gives the batch weight formulation for the production of 260 l of paint.

Liquid ingredients (the resin, solvent, dispersing agent, biocide and water) are usually added by volume (in litres) and this is obtained by dividing

Table 3.2 Batch Formulation of a White Emulsion Paint

	Weight (%)	Weight* (kg)	Volume† (l)
Rutile titanium dioxide	19·2	70·66	—
Whiting	21·7	79·86	—
Emulsion resin	27·9	102·67	100·66
Thickener	0·7	2·58	—
Dispersing agent	0·5	1·84	1·84
Biocide	0·3	1·10	1·10
Coalescing solvent	0·5	1·84	1·88
Water	29·2	107·46	107·46

*Weight per cent × 368/100
†Weight (kg) divided by density (kg/l)

Note: Both weights and volumes are normally rounded off to facilitate actual paint production.

the mass (weight in kg) by its density (kg/l). The volume additions are given in Table 3.2 but both the weight and volumes are normally rounded off for convenience during paint manufacture.

The batch formulation given in Table 3.2 is used when calculating raw materials costs since the latter are usually quoted as pounds sterling (or dollars) per kg or litre, or multiples of these basic units. Thus if the emulsion resin cost £280 per 1000 kg, the cost of the 102·67 kg used in the paint would be

$$\pounds\ \frac{102{\cdot}67}{1000}\times 280=\pounds 28{\cdot}75$$

3.2 Specific gravity

In the above calculations, specific gravities (or densities) were known or could be reasonably assumed. In many instances, however, this is not the case and the specific gravity has to be calculated for the particular component in question.

3.2.1 CUMULATIVE SPECIFIC GRAVITIES

Blends of resins or resin–plasticiser combinations are used in many paint formulations and the cumulative specific gravity must be calculated in order to derive the weight or volume addition.

Example B

In a chlorinated rubber–plasticiser combination, the rubber has s.g. 1·5 while the chlorinated paraffin plasticiser has s.g. 1·15. The resin–plasticiser combination is used at a weight ratio of 0·6:1·0 and the composite specific gravity is required.

Firstly, the weight ratio of 0·6:1·0 (resin:plasticiser) is converted to a percentage ratio:

$$\text{resin} \qquad \frac{0{\cdot}6}{1{\cdot}6}=0{\cdot}375$$

$$\text{plasticiser} \qquad \frac{1{\cdot}0}{1{\cdot}6}=0{\cdot}625$$

The total volume is calculated using the relation volume $= \dfrac{\text{mass}}{\text{specific gravity}}$:

	% by weight	*s.g.*	*volume*
resin	37·5	1·5	25·0
plasticiser	62·5	1·15	54·35
total	100		79·35

The cumulative or combined s.g. of the blend, 1·26, is then obtained by dividing the total weight by the total volume:

$$\text{s.g. } \frac{100}{79{\cdot}35}=1{\cdot}26$$

3.2.2 SPECIFIC GRAVITY OF A POLYMER IN SOLUTION

The formulation given in Table 3.1 is based on an aqueous emulsion resin containing 52 per cent by weight of solids, s.g. being 1·02. The specific gravity of the solid polymer is obtained as follows.

Example C

Combined s.g.=1·02, the solid resin providing the 'additional' 0·02 s.g. Consequently,

$$\text{s.g. of solid resin} = \left(\frac{100}{52}\times 0{\cdot}02\right)+1{\cdot}0000$$

$$= (1{\cdot}923\times 0{\cdot}02)+1{\cdot}0000$$

$$= 1{\cdot}0385$$

A slightly different approach is required when the specific gravity of the polymer solution is less than unity.

Example D

A solution of alkyd resin in white spirit contains 75 per cent by weight of resin. The s.g. of white spirit is 0·80 while the solution s.g. is 0·97.

The proportionate specific gravity of the white spirit solvent is (25/100)×0·80=0·20. This value is deducted from the solution s.g. to give the contribution from the alkyd resin:

$$0{\cdot}97-0{\cdot}20=0{\cdot}77$$

From this value, the s.g. of the alkyd resin is calculated:

$$\frac{100}{75}\times 0{\cdot}77=1{\cdot}0266$$

In fact, this approach can also be used for solutions of specific gravity greater than 1·0000 but the first method given is somewhat simpler and quicker.

3.3 Drier additions

Paints based on oxidation-curing media such as oleoresinous and alkyd binders require the addition of driers (heavy metal salts of organic acids, page 30) to effect a reasonable curing rate. The level of drier addition (proportion of metal to solid resin) is either specified by the resin manufacturer or determined by the proposed service application of the paint. Consequently, drier additions have to be made with some care to achieve the required effect.

Example E

An alkyd resin requires 0·5 per cent lead metal and 0·06 per cent cobalt metal driers based on the solids content of the binder solution. Drier solutions, as supplied by the manufacturer, are specified by the metal content in solution, i.e. 24 per cent lead and 6 per cent cobalt when supplied as an organic acid salt such as the naphthenate.

The amount of lead naphthenate solution, containing 24 per cent lead metal, required to give 0·5 per cent lead metal for, say, 75 kg of alkyd resin is

$$\frac{75}{100} \times 0{\cdot}5 = 0{\cdot}375 \text{ kg lead metal}$$

The lead naphthenate solution contains 24 per cent lead so that $(100/24) \times 0{\cdot}375 = 1{\cdot}56$ kg solution will provide 0·375 kg lead metal.

Similarly, the amount of cobalt metal (at an addition level of 0·06 per cent) required for 75 kg alkyd resin is

$$\frac{75}{100} \times 0{\cdot}06 = 0{\cdot}045 \text{ kg}$$

The weight of 6 per cent cobalt naphthenate solution required to yield 0·045 kg cobalt metal is

$$\frac{100}{6} \times 0{\cdot}045 = 0{\cdot}75 \text{ kg}$$

3.4 Percentage volume solids

An important aspect of paint technology is the calculation of application rates (that is, covering capacity) and film thicknesses, for which it is necessary to know the solids content of the paint formulation. Again, example calculations are based on the formulation given in Tables 3.1 and 3.2.

The formulation was given in terms of percentage weights of the components and these values should be converted to percentage weight solids contents. The rutile titanium dioxide and whiting are obviously 100 per cent solids and present no problems. For convenience, the cellulosic thickener can also be assumed to be 100 per cent solid. The water and coalescing solvent are obviously 100 per cent liquid. The other components (emulsion resin, dispersing agent and biocide) are solids dispersed in a volatile fluid phase. It is assumed here that the solids contents of the dispersing agent and biocide are small (less than 10 per cent) so that in relation to the total solids content their contribution can be neglected and they can be regarded as volatiles.

Example F
The emulsion resin has a solids content of 52 per cent by weight and therefore, for a 29 per cent weight addition to the formulation, the weight percentage of solid resin is

$$\frac{52}{100}\times 27{\cdot}9 = 14{\cdot}51\%$$

The specific gravity of the resin, as determined in Section 3.2.2 above, is 1·0385 and can be rounded off to 1·04.

The percentage volume solids for each component is then obtained by dividing its weight by its specific gravity to obtain the volume: volume=mass/s.g. These values are given in Table 3.3.

The percentage volume solids content of the paint may then be derived using the relation

$$\text{percentage volume solids content} = \frac{\text{total volume solids}}{\text{total volume}}\times\frac{100}{1}\%$$

$$= \frac{26{\cdot}80}{70{\cdot}71}\times\frac{100}{1}\%$$

$$= 37{\cdot}90\%$$

which can be rounded off to 38 per cent.

Table 3.3 Percentage Volume Solids

	Weight (%)	s.g.	Volume (total)	Solids (%)	s.g.	Volume (solids)
Rutile TiO_2	19·2	4·2	4·57	19·2	4·2	4·57
Whiting	21·7	2·8	7·75	21·7	2·8	7·75
Emulsion resin	27·9	1·02	27·35	14·51	1·04	13·95
Thickener	0·7	1·33	0·53	0·7	1·33	0·53
Dispersing agent	0·5	1·0	0·50	—	—	—
Biocide	0·3	1·0	0·30	—	—	—
Coalescing solvent	0·5	0·98	0·51	—	—	—
Water	29·2	1·0	29·20	—	—	—
Totals	100	—	70·71	56·11	—	26·80

3.5 Paint application rate and film thickness

The percentage volume figures are required for the calculation of the application rate, this being the theoretical coverage, in m^2 per litre of paint, achieved with a given formulation at a particular dry paint film thickness. The relationship for the application rate or surface coverage is

$$\text{theoretical coverage } (\text{m}^2/\text{l}) = \frac{\text{percentage volume solids}}{\text{dry film thickness } (\mu\text{m})} \times 10$$

Example G

For the white emulsion paint used here, the percentage volume solids is 38 per cent and, for a 40 μm thick dry film,

$$\text{application rate} = \frac{38}{40} \times 10 = 9{\cdot}5 \text{ m}^2/\text{l of paint}$$

The surface area covered by unit volume of paint (i.e. the application rate) can be obtained from the expression

$$\text{application rate } (\text{m}^2/\text{l}) = \frac{\text{weight per litre of paint}}{\text{weight of liquid paint applied to 1 m}^2}$$

The weight of one litre of paint is best determined experimentally but a theoretical value can be obtained by the weight percentage divided by the volume of paint.

Example H

Weight percentage of paint is 100 while the volume is 70·71 (Table 3.1); thus

$$\text{weight per litre} = \frac{100}{70{\cdot}71} = 1{\cdot}414 \text{ kg/l}$$

The wet film thickness can be calculated from the percentage volume solids and dry film thickness:

$$\text{wet film thickness } (\mu\text{m}) = \frac{\text{dry film thickness}}{\%\text{ volume solids}} \times 100$$

Example I

$$\text{Wet film thickness} = \frac{40}{38} \times 100 = 105{\cdot}3\ \mu\text{m}$$

The wet film thickness can also be calculated using the relationship

$$\text{wet film thickness } (\mu\text{m}) = \frac{\text{volume of paint applied (cm}^3)}{\text{area covered by the paint (m}^2)}$$

Example J

$$\text{Wet film thickness} = \frac{1000}{9{\cdot}5} = 105{\cdot}3\ \mu\text{m}$$

3.6 Paint requirements

In test laboratories it is standard procedure to convert theoretical coverage or application rate (m^2/l) into the weight of paint to be applied to a test panel of specified dimensions. The weight of paint applied is given by

$$\text{weight of paint (g)} = \frac{\text{weight per litre of paint (g)}}{\text{application rate (cm}^2\text{/l)}} \times \text{panel area (cm}^2\text{)}$$

Example K
The paint used here has a weight per unit volume of 1·414 kg/l and an application rate of 9·5 m^2/l at a dry film thickness of 40 μm. Thus, when applied to a test panel 10×15 cm (150 cm^2) at a rate of 9·5 m^2/l,

$$\text{weight of paint} = \frac{1{\cdot}414}{95\ 000} \times 150 = 2{\cdot}23 \text{ g}$$

A figure often required in practice is the volume of paint required to cover a specified surface area at a given dry film thickness. This can be obtained using the relationship

$$\text{volume of paint (l)} = \frac{\text{area (m}^2) \times \text{dry film thickness } (\mu\text{m}) \times 10}{\%\ \text{volume solids} \times 100}$$

Example L
Using the white emulsion paint (38 per cent volume solids), the volume of paint required to provide a dry film thickness of 60 μm over 500 m^2 of wall is given by

$$\text{volume of paint} = \frac{500 \times 60 \times 10}{38 \times 100}$$

$$= \frac{300\ 000}{3800} = 78{\cdot}9 \text{ litres}$$

which can be rounded off to 79 litres.

3.7 Pigment-to-binder ratio

In previous sections, an example of a white emulsion paint was used to demonstrate the interconversion of paint formulae and the calculation of application rates, wet film thicknesses and paint requirements. It was assumed that the formulation was optimal for the end use (service application) of the paint although the latter was not specified nor was there any discussion of how the formulation was actually derived. The formulation of a paint for a particular coating application is a complex procedure and is discussed in other chapters. However, two parameters are important in paint formulation, namely the pigment:binder ratio and the pigment volume concentration.

The pigment:binder ratio of a paint is obtained from the percentage weight content of the pigmentation (that is, pigments and extenders) expressed as a weight ratio of the solid (non-volatile) resin percentage weight content.

Example M

In the paint example (Table 3.1), the total pigment and extender content (titanium dioxide and whiting) is 19·2+21·7=40·9 per cent by weight. The solid resin content was found to be 14·51 per cent by weight (Section 3.4). Consequently, the percentage weight ratio of pigmentation to resin solids is 40·9:14·51 or 2·8:1·0.

Paints generally fall into fairly well-defined categories of pigment:binder ratios and this is a useful guide during the initial formulation of a paint. On the other hand, it is very difficult to predict (or even calculate) the total solids content to be used in a paint formulation since this value is determined by a number of considerations (see Chapter 2). The total solids requirements for a particular application can normally be estimated from established paint formulations that have acceptable characteristics.

3.8 Pigment volume concentration

In previous sections of this chapter, paint formulations have been expressed in terms of weight. However, the resins, pigments, extenders and other solid additives can differ widely in their specific gravities. Consequently, their volumes or, better, volume contributions to the overall film are markedly different. This is of greater importance than the weight percentages in determining the properties of the dry film and is particularly important when undertaking comparative performance tests on differently formulated paints.

The volume relationship between the solids in the film (resin binder, pigments and extenders) is expressed as the pigment volume concentration (P.V.C.):

$$\text{P.V.C.} = \frac{\text{volume of pigments and extenders}}{\text{volume of pigments and extenders} + \text{volume of solid binder}} \times \frac{100}{1}\%$$

The volume of the pigments and extenders is obtained from their percentage addition weights in the formulation using the relation: volume =mass/s.g. The specific gravities may be readily available in the literature or from the manufacturer, or they may be determined experimentally (see Section 3.8.1).

In order to calculate the P.V.C. of a formulation, the percentage weights and specific gravities of all the solid components are required.

Example N

Using the values given for the above parameters in Table 3.3, the volumes of the three major constituents are the following:

rutile titanium dioxide	4·57
whiting	7·75
emulsion resin	13·95

Consequently, the P.V.C. is given by

$$\text{P.V.C.} = \frac{4{\cdot}57+7{\cdot}75}{4{\cdot}57+7{\cdot}75+13{\cdot}95} \times \frac{100}{1}\%$$

$$= \frac{12{\cdot}32}{26{\cdot}27} \times \frac{100}{1}\% = 46{\cdot}9\%$$

The significance in paint formulation of pigment volume concentration (P.V.C.) and critical pigment volume concentration (C.P.V.C.) has been discussed in Chapter 2.

3.8.1 SPECIFIC GRAVITY OF PIGMENTS

If the specific gravity of a pigment or extender is not known, it can be determined using a density bottle. This is a specially made vessel of accurately known internal volume, typically 100 cm³.

Specific gravity measurements are made as follows: the bottle is weighed empty (weight w_1) and then reweighed when partially filled with the pigment (weight w_2). The bottle is then filled with an organic solvent, care being taken to expel any entrapped air, and reweighed (weight w_3). Finally, the bottle is emptied, cleaned, filled with clean solvent and reweighed (weight w_4).

Thus, for a 100 cm³ density bottle,

$$\text{specific gravity} = \frac{k \times (w_2 - w_1)}{(w_4 - w_1) - (w_3 - w_2)}$$

where w_1 = weight of density bottle
w_2 = bottle + pigment
w_3 = bottle + pigment + solvent
w_4 = bottle + solvent
k = s.g. of solvent.

3.8.2 OIL ABSORPTION

Knowledge of the oil-absorption value of a pigment or extender is a prerequisite in paint formulation; this value is also commonly obtained as a raw material quality control procedure. The oil-absorption value can be defined as the amount of binder (traditionally acid-refined linseed oil having an acid value of 7·5–8·5 mg KOH per gramme) that is required to convert a given weight of the powder under test to a paste.

Basically, the oil is volumetrically added to a weighed quantity of the powder which is vigorously blended using a spatula until a just coherent paste is achieved. The oil-absorption value (O.A.) of the powder is usually expressed as the weight of oil required to convert 100 grammes of the material under test into a paste. This can be calculated from

$$\text{O.A. (wt/wt)} = \frac{(\text{s.g. of binder}) \times (\text{volume (ml) of binder required})}{\text{weight (g) of pigment taken}}$$

The oil-absorption value of a pigment or extender is controlled by a number of important factors, not least of which is the efficacy of the operator performing the test. These include the shape and size of the powder particles, as well as the size distribution. Other factors are the presence of absorbed water which can affect the wettability of the pigment, as well as the wetting characteristics of the binder used.

Using oil-absorption data, it is possible to obtain a value for the critical pigment volume concentration of a formulation (pages 37 and 41), although clearly this can only be a guide if binder types other than linseed oil are used in the particular paint. The critical pigment volume concentration can be determined from oil-absorption data from the formula

$$\%\ \text{C.P.V.C.} = \frac{1}{1+\text{O.A.}} \times \frac{100}{1}$$

3.9 Basic statistics

3.9.1 INTRODUCTION

In Sections 3.1 to 3.8, the essential mathematics of paint technology has been outlined. There is, however, another important series of mathematical operations that should be involved in virtually all aspects of paint technology, namely statistical analysis. Clearly it is not possible to provide a detailed discussion of statistics in this book, but a brief guide to simple statistical analysis may be useful.

A statistic is an item of information, usually numerical, and the study of data using such items of information is known as the science of statistics.

Although there tends to be a high degree of scepticism with regard to statistics, often justifiable in view of their abuse or misuse by the media, advertising campaigns and politicians, they do represent the most convenient method of recording data. Statistical analysis is, in fact, an exact science but it can be misapplied and this is the problem central to doubts about the accuracy or validity of statistics.

In this chapter a basic outline of statistics is given but there is little explanation of the underlying theory, which is more correctly the province of the specialised textbook (see the bibliography); it would clearly be impossible to discuss statistical theory here. Consequently, the approach adopted is to present the statistical test and provide an example rather than to justify all the statements made.

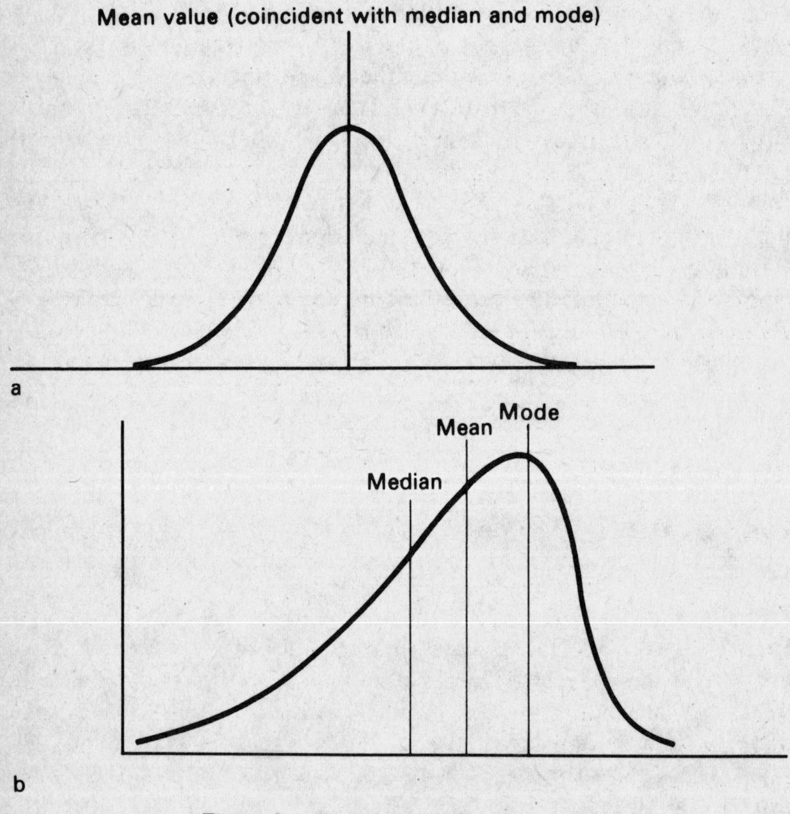

Figure 3.1 Distribution of data
a Symmetrical distribution
b Skewed distribution

3.9.2 AVERAGES AND MEANS

The average value, meaning the typical value, is the most widely used statistic but in fact several 'averages' exist. The one usually referred to is the mean, or better the arithmetic mean, defined as

$$\text{arithmetic mean} = \frac{\text{sum of values}}{\text{number of values}} = \frac{\Sigma x}{n}$$

and it has the symbol $\bar{x}$ for a series of measurements of some property x.

Example O
If there are seven measurements of x, i.e. $n=7$, and the x values obtained are 4, 6, 6, 7, 8, 9, 10,

$$\text{arithmetic mean, } \bar{x} = \frac{4+6+6+7+8+9+10}{7} = 7{\cdot}14$$

The mean value is a measure of central tendency in that it tends to lie centrally or in the middle of the data. In contrast, the difference between the highest and lowest figures (10−4=6) is the range of the values.

Two other measures of central tendency are sometimes encountered, the mode and the median. The mode (or modal value) is the most commonly occurring valuc in a group of data, i.c. 6 in the above example since it is the one figure that occurs twice. The median value is the middle value of the series of data, and the median is 7 in the example. In many instances, the mean, mode and median will all coincide, as in a symmetrical distribution of data (see later), but not always. A symmetrical distribution, known as the normal curve, has a bell-shape, the peak of which is the mean value (Figure 3.1a). Such curves often relate to, say, measurements of heights, weights, IQ values, etc., and are very important in statistics. Non-symmetrical or skewed distributions do not have the peak or mean value at the centre of the curve (Figure 3.1b) and can relate to data such as plots of production figures, e.g. greater output towards the end rather than at the start of the week.

3.9.3 CALCULATION OF MEANS

Obtaining the mean for a data sample is straightforward and involves summing all the values and dividing the result by the number of items. It is fairly common, however, for a particular value to occur more than once in a set of data. In such cases, the mean is given by

$$\text{mean, } \bar{x} = \frac{\Sigma f_i x_i}{n}$$

where f_i is the frequency (number of times) that value x_i occurs.

Example P
Data: 5,5,6,6,6,6,7,7,7,7,7,8,8,9,9,9,9,10,10,10

$$\text{Mean, } \bar{x} = \frac{(2\times5)+(4\times6)+(5\times7)+(2\times8)+(4\times9)+(3\times10)}{20}$$

$$= \frac{151}{20} = 7{\cdot}55$$

3.9.4 VARIATION

The mean, mode and median are measures of the central tendency of data while the range is an indication of the variation of the data. However, often it is necessary to know the deviation of individual values from the mean, i.e. the scatter of the data; this is important when comparing data (see later).

There are two measures for the scatter of data, the mean deviation and the standard deviation; the latter, denoted by the symbol s, is the more commonly used statistic. The square of the standard deviation is the variance, s^2, and this is also an important statistic.

The standard deviation is defined by the expression

$$\text{standard deviation, } s = \sqrt{\left(\frac{\Sigma(x-\bar{x})^2}{n-1}\right)}$$

where $\bar{x}$ is the mean value for the data, x is an individual item of information and n is the number of items of information. Clearly, it would be tedious to use the above expression to calculate the standard deviation, s, and a 'short-cut' approach is usually adopted:

$$s = \sqrt{\left(\frac{n(\Sigma\, x^2)-(\Sigma x)^2}{n(n-1)}\right)}$$

To calculate s, both the sum of the individual values (Σx) and that of the squares of the individual values (Σx^2) are obtained and then the calculation is performed.

Example Q
If the data comprise $x=1, 2, 3, 4, 5$

$$\Sigma x=15, \quad (\Sigma x)^2=225, \quad \Sigma x^2=55, \quad n=5, \quad n(n-1)=20$$

Hence

$$\bar{x} = \frac{\Sigma x}{n} = \frac{15}{5} = 3{\cdot}0$$

$$s = \sqrt{\left(\frac{(5\times55)-225}{20}\right)} = \sqrt{\left(\frac{50}{20}\right)} = 1{\cdot}5811$$

When expressing mean values, it is common to give both the mean value and the standard deviation. In the above example, $\bar{x}\pm x=3{\cdot}0\pm1{\cdot}58$.

The standard deviation is an absolute measure of the dispersion or scatter of data and cannot always be used to compare two sets of data. For this, one can use the relative dispersion or coefficient of variation, V:

$$\text{coefficient of variation, } V=\frac{\text{absolute dispersion}}{\text{mean}}=\frac{100}{1}\%$$

$$=\frac{s}{\bar{x}}\times\frac{100}{1}\%$$

Example R

Two manufacturing processes, A and B, have the same standard deviations but different mean values: Process A: 100±10 gallons/day; Process B: 500 ±10 gallons/day.

$$\text{Process A:}\quad V_{\mathrm{A}}=\frac{10}{100}\times\frac{100}{1}\%=10\%$$

$$\text{Process B:}\quad V_{\mathrm{B}}=\frac{10}{500}\times\frac{100}{1}\%=2\%$$

It can be seen that Process A is the more variable of the two.

Finally, each individual value x can be expressed in terms of deviation from the mean value $\bar{x}$ in units of the standard deviation, s. This value of x is known as the standard score, denoted by Z:

$$\text{standard score, } Z=\frac{x-\bar{x}}{s}$$

Example S

In Example Q above, $\bar{x}=3$ and $s=1{\cdot}58$, so that the standard score for, say, 5 is given by

$$Z=\frac{5-3}{1{\cdot}58}=1{\cdot}27$$

while for 2,

$$Z=\frac{2-3}{1{\cdot}58}=-0{\cdot}63$$

Standard scores are important when comparing data (see later) but an example here may be useful.

Example T

In a series of tests on one day, product A gave a strength figure 11 MN/m^2 while the overall mean and standard deviation for the range of products tested was $10 \pm 2{\cdot}12$ MN/m^2. On the following day, testing the same products with the same test regimen, product A gave a figure of 12 MN/m^2 while for all the products, $\bar{x}=10{\cdot}8$ and $s=1{\cdot}48$.

Day 1:

$$\text{standard score for A, } Z_A = \frac{11-10}{2{\cdot}12} = 0{\cdot}47$$

Day 2:

$$\text{standard score for A, } Z_A = \frac{12-10{\cdot}8}{1{\cdot}48} = 0{\cdot}81$$

It can be seen that product A had a relatively better strength than the 'average' on day 2 than on day 1.

3.9.5 NORMAL CURVE AND PROBABILITIES

The curve shown in Figure 3.1a has a characteristic shape and is known as the *normal curve*. All data from a series of measurements will be enclosed by (or are 'under') this curve and the area under it by definition is unity. This means that the probability, P, of finding the value of an individual measurement under the curve is 100 per cent, i.e. $P=1{\cdot}0$.

In discussing the normal curve, one distinguishes between a *population* (a very large series of measurements or the main source used for sampling, e.g. the total number of men in the UK) and a *sample* (a limited number of measurements, e.g. a random selection of men for a height survey). For a sample, the mean and standard deviation are denoted by the symbols $\bar{x}$ and s respectively, while the mean and standard deviation of a population are given the symbols μ and σ (Figure 3.2).

The peak of the normal curve lies at the mean (μ) value while the area of the curve is taken to lie within the limits $\mu+4\sigma$ and $\mu-4\sigma$, that is, within four standard deviations either side of the mean. The area lying between $\mu-\sigma$ and $\mu+\sigma$, that is, within one standard deviation either side of the mean, is 68·26 per cent of the total area. Similarly 95·45 per cent and 99·73 per cent of the area lie between $\mu \pm 2\sigma$ and $\mu \pm 3\sigma$, respectively. Thus the probability of finding any value within the range $\mu \pm \sigma$ is 68·26 per cent, or $P=0{\cdot}6826$.

Probabilities underlie the whole science of statistics and range from $P=1{\cdot}0$, absolute certainty, to $P=0{\cdot}0$, completely impossible. They are used in statistical tests (see later).

The standard score, Z, can be used to transform all normal curves, that is, curves with different values for μ and σ, into a standard or unit normal curve. The standard score Z was defined earlier as

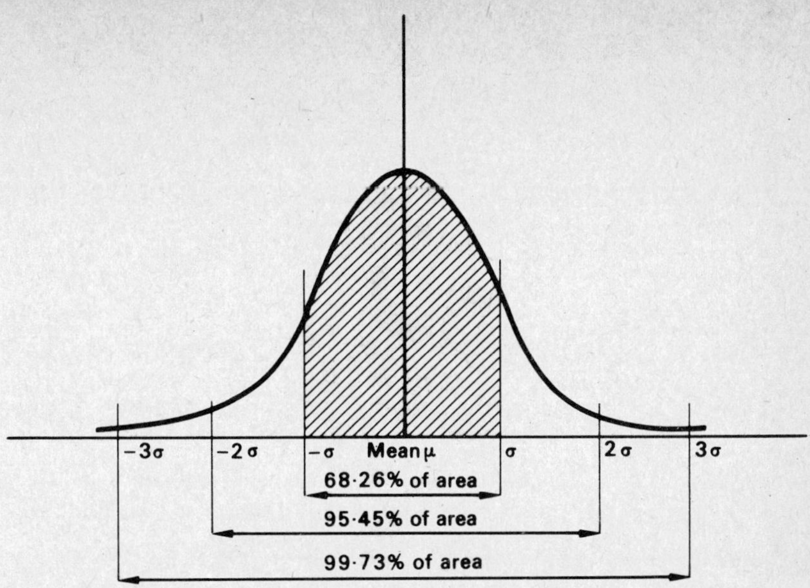

Figure 3.2 Areas under the normal curve (total area = 1·0)

$$Z=\frac{x-\mu}{\sigma}$$

so that $Z=0$ at the mean value μ, $Z=\pm1$ at $\mu\pm\sigma$, $Z=\pm2$ at $\mu\pm2\sigma$, etc. Thus the area under the curve is encompassed by $Z=-4$ to $Z=4$. Furthermore, the area between any two Z values represents thc probability of the measurement lying within that range. Tables giving the area under the normal curve for different Z values are readily available.

Example U

An example of the use of probabilities is in expected values. If the probability of an event is 0·5 (e.g. tossing a coin gives 'heads' or 'tails', so that $P=0{\cdot}5$ for heads, and 0·5 for tails, since both are equally possible), then in performing a trial 50 times ($n=50$), the event should occur $50\times0{\cdot}5=25$ times:

$$\begin{aligned}\text{expectation} &= n\times P\\ &= 50\times0{\cdot}5=25\end{aligned}$$

Alternatively, if the production figures for a process were 260±25 gallons per day, it might be of interest to know the probability of production of 300 gallons or greater, and how often this might occur over a period of 200 working days.

The standard score for 300 gallons is given by

$$Z=\frac{300-260}{25}=1{\cdot}6$$

and the area of the normal curve corresponding to $Z\geq1{\cdot}6$ (i.e. 300+ gallons) is 0·055 (from standard tables), so that the probability of a daily production figure of 300 or more gallons is 0·055. The expectation or number of times this figure would be met or exceeded in 200 working days is

$$E=200\times0{\cdot}055=11$$

that is, it would be expected to occur 11 times in this period or roughly once every twenty days. Conversely, a production of 220 ($Z=-1{\cdot}6$) could also occur 11 times during the same period since both events are equally probable.

3.10 Statistical tests

3.10.1 INTRODUCTION

In undertaking a statistical test, the implied question is either (a) 'Do the two series of measurements actually differ?' or 'Is the difference in mean values statistically significant?' or (b) 'Does the mean value obtained for a sample differ significantly from that of the population from which it was drawn?'

The usual probability levels for statistical significance are:

Probably significant:	$P<0{\cdot}05$	(probability of less than 1 in 20 or 5 per cent)
Significant:	$P<0{\cdot}01$	(probability of less than 1 in 100 or 1 per cent)
Highly significant:	$P<0{\cdot}001$	(probability of less than 1 in 1000 or 0·1 per cent)

A probability of greater than 1 in 20, i.e. $P>0{\cdot}05$, is usually taken to be non-significant, while statistical significance is only assumed if $P<0{\cdot}01$. If $P<0{\cdot}05$ but $P>0{\cdot}01$, the difference is only probably significant and further confirmatory measurements should be undertaken.

The above probabilities representing significance levels of 5 per cent, 1 per cent and 0·1 per cent are of great importance in statistical tests. There are critical values of the Z score, Z_c values, corresponding to these significance levels (Table 3.4). The one-tail test Z_c values are used (see later) when one is concerned with whether the measured value is greater than (or smaller than) the comparative data, while the two-tail Z_c values are used if

Table 3.4 Critical Z Values (Z_c)

	Significance level		
	5%	1%	0·1%
One tail test	1·64	2·33	3·08
Two-tail test	1·96	2·58	3·27

the question is whether the measured values differ from the population mean. The terms 'one-tail' and 'two-tail' refer to the probability of the standard score (Z value) of the test data lying in the area of the curve (the 'tail') corresponding to that significance level.

3.10.2 STANDARD ERROR OF THE MEAN

The standard error of the mean, $\bar{x}$, for a sample consisting of n measurements and a standard deviation s is given by:

$$SE(\bar{x}) = \frac{s}{\sqrt{n}}$$

The standard error is a measure of the precision with which the population mean, μ, can be estimated from the sample mean, $\bar{x}$. It is important because the means are obtained for several batch samples from a population and so the individual sample means and standard deviations will differ slightly from each other. These several separate means and standard deviations will themselves have a mean and standard deviation: this is known as a sampling distribution of the means, or standard error of the means, and the general expression for this is:

$$SE(\mu) = \frac{\sigma}{\sqrt{n}}$$

for a population of mean μ and standard deviation σ.

This statistic can be used to test whether a test or sample mean, $\bar{x}$, differs significantly from the population mean, μ. For this, the standard score Z is calculated as

$$Z = \frac{\bar{x} - \mu}{SE\,(\mu)}$$

Example V

Consider a paint formulation with a specified mean viscosity of 0·5 N s m^{-2} and a standard deviation of 0·05, (i.e. the population viscosity is 0·5±0·05 N s m^{-2}). The viscosities of two batches of this paint, based on a series of 10 determinations each, are 0·57±0·07 and 0·48±0·08 N s m^{-2}, respectively. Do the batch viscosities differ significantly from the mean value of 0·50?

Batch A: $$SE(\mu)=\frac{s}{\sqrt{n}}=\frac{0{\cdot}07}{\sqrt{10}}=0{\cdot}022$$

Batch B: $$SE(\mu)=\frac{s}{\sqrt{n}}=\frac{0{\cdot}08}{\sqrt{10}}=0{\cdot}025$$

Thus

$$Z_A=\frac{\bar{x}-\mu}{SE(\mu)}=\frac{0{\cdot}57-0{\cdot}50}{0{\cdot}022}=3{\cdot}18$$

and

$$Z_B=\frac{0{\cdot}48-0{\cdot}50}{0{\cdot}025}=-0{\cdot}80$$

In the case of Batch A, $Z_A=3{\cdot}18$ which is greater than Z_C for a one-tail test at the 0·1 per cent significance level (see Table 3.4), and the viscosity of this batch is highly significantly ($P<0{\cdot}001$) greater than that of the population (specification) mean value. With Batch B, $Z_B=-0{\cdot}80$ and, although the batch viscosity appears to be lower than that for the population, this difference is non-significant ($P>0{\cdot}05$) since Z_C for the 5 per cent level is −1·64.

This was a straightforward statistical test because there was a population mean (the specification viscosity) with which to compare the sample values. In many instances, however, this is not the case, although one may still wish to test whether the difference in mean values for two batches ($\bar{x}_1$ and $\bar{x}_2$) is statistically significant. For this the standard error of the difference between means, $SE(\bar{x}_1-\bar{x}_2)$, is used. This statistic is defined as follows:

$$SE(\bar{x}_1-\bar{x}_2)=\sqrt{\left(\frac{(S_1)^2}{n_1}+\frac{(S_2)^2}{n_2}\right)}=\sqrt{\{[SE(\bar{x}_1)]^2+[SE(\bar{x}_2)]^2\}}$$

where n_1, S_1 and n_2, S_2 are the sample sizes and standard deviations for the two samples.

If the difference between the means is greater than twice the standard error, i.e. $(\bar{x}_1-\bar{x}_2)>2\,SE(\bar{x}_1-\bar{x}_2)$, then the difference is probably significant ($P<0{\cdot}05$). If the difference is greater than three times the standard error, the difference is statistically significant ($P<0{\cdot}01$).

Example W
In the case of the two batch viscosities (Example V), the difference in mean values $\bar{x}_1 - \bar{x}_2 = 0{\cdot}57 - 0{\cdot}48 = 0{\cdot}09$.

$$SE(\bar{x}_1 - \bar{x}_2) = \sqrt{\left(\frac{(0{\cdot}07)^2}{10} + \frac{(0{\cdot}08)^2}{10}\right)} = 0{\cdot}034$$

Thus

$$(\bar{x}_1 - \bar{x}_2) = 2{\cdot}65 \times SE\ (\bar{x}_1 - \bar{x}_2)$$

and the difference is probably significant but not definitely significant, i.e. $0{\cdot}01 < P < 0{\cdot}05$. The result suggests that further tests should be performed to determine whether a real (statistically significant) difference exists.

This test is rapid and convenient but is somewhat imprecise. A more sophisticated approach, namely Student's *t* test, would give more precise information.

3.10.3 STUDENT'S *t* TEST

When comparing the mean of a sample of *n* values with that of a population, the standard score, *Z*, can be used if the standard deviation σ of the population is known. If this is not the case, as is often so, it is possible to estimate the population variance (σ^2) from the sample variance (S^2) and calculate the *t* statistic:

$$t = \frac{\text{difference between means}}{\text{estimated standard error of the mean}}$$

$$= \frac{(\bar{x} - \mu)}{S^*} \times \sqrt{n}$$

where the estimated standard deviation of the population, S^*, is given by

$$S^* = \sqrt{\left(\frac{\Sigma\ (x - \bar{x})^2}{n-1}\right)}$$

The *t* statistic is used in a similar way to the *Z* value except that tables of values for the *t* statistic (the *t* distribution) are used instead of the normal distribution as in the case of *Z*. There is also one other factor that has to be taken into account and that is the degrees of freedom (*df*). This is the total number of measurements, less one; thus, if the means of two sets of measurements ($n_1 = 9$, $n_2 = 11$) are being compared, the degrees of freedom for the *t* statistic are

$$df = (n_1 - 1) + (n_2 - 1) = (9 + 11) - 2 = 18$$

The value of the t statistic changes with the df, as can be seen from any set of statistical tables, and in determining the statistical significance of a calculated t value, that is, whether it exceeds the critical t value for the significance level adopted (analogous to the Z_c levels given in Table 3.4), the correct df must be used.

Example X

Considering the paint viscosities in Example V, it is questioned whether the difference between the two batch viscosities is due to sampling variation or is a real difference (due to changes in formulation or processing).

For two groups n_1, n_2 with means and standard deviations $\bar{x}_1 \pm S_1$, $\bar{x}_2 \pm S_2$,

$$t=\frac{\bar{x}_1-\bar{x}_2}{\sqrt{\{S^2[(1/n_1)+(1/n_2)]\}}}$$

where S^2, the pooled variance, is given by

$$S^2=\frac{(n_1-1)S_1^2+(n_2-1)S_2^2}{(n_1+n_2)-2}$$

and the degrees of freedom of the calculated t value,

$$df=(n_1-1)+(n_2-1)=(n_1+n_2)-2$$

Batch A: $n_1=10$, $\bar{x}_1=0{\cdot}57$, $S_1=0{\cdot}07$
Batch B: $n_2=10$, $\bar{x}_2=0{\cdot}48$, $S_2=0{\cdot}08$

Now

$$S^2=\frac{(10-1)\,(0{\cdot}07)^2+(10-1)\,(0{\cdot}08)^2}{10+10-2}=5{\cdot}63\times10^{-3}$$

and $S=0{\cdot}075$.

$$t=\frac{\bar{x}_1-\bar{x}_2}{\sqrt{\{S^2[(1/n_1)+(1/n_2)]\}}}$$

$$=\frac{0{\cdot}57-0{\cdot}48}{\sqrt{\{5{\cdot}63\times10^{-3}[(1/10)+(1/10)]\}}}$$

$$=\frac{0{\cdot}09}{0{\cdot}034}=2{\cdot}647 \quad \text{with } df=18$$

From tabulated t values, the critical value of t for the 1 per cent probability level is 2·552 when $df=18$. Thus the viscosity of Batch A is significantly higher ($P<0{\cdot}01$) than that of Batch B.

In performing the t test, it is assumed that the means are identical, and then the probability is tested that the two means could have come from the same population. Secondly, it is also assumed that the two variances are identical, since if there is a significant difference between them, the t test results have to be treated with caution.

Logically, therefore, the variances should be compared before undertaking a t test. Although this is often neglected, it is in fact a prerequisite of a valid t test. The variances are compared using the variance ratio or F test:

$$F=\frac{\text{greater estimate of population variance}}{\text{lesser estimate of population variance}}$$

The estimate of the variance is the sum of the squares of the deviations from the mean divided by $(n-1)$:

$$\text{variance estimate}=\frac{\Sigma(x-\bar{x})^2}{n-1}$$

For large samples the variance ratio F will be close to unity but it can differ markedly for small n values. Tabulated F values are available for different degrees of freedom (the $(n-1)$ values) at various significance levels, usually 5 per cent and 1 per cent, and the calculated F value is compared with the tabulated or theoretical value. If the calculated value exceeds that given in the table, then there is a significant difference in the variances.

Example Y

In Example V, the viscosity variance (S^2) of Batch A was $4{\cdot}9\times10^{-3}$, while that of Batch B was $6{\cdot}4\times10^{-3}$, so that

$$F=\frac{6{\cdot}4\times10^{-3}}{4{\cdot}9\times10^{-3}}=1{\cdot}306$$

However, since the theoretical F value (at the 5 per cent level) for the two degrees of freedom concerned, $(n_1-1)=(n_2-1)=9$, is 3·23, the two variances do not differ significantly and a t test can be performed.

3.10.4 ANALYSIS OF VARIANCE

Analysis of variance is a method of examining the effect of one or more factors on the sample data. A one-way analysis of variance examines the effect of one factor, two-way analysis examines two factors and a three-way analysis examines the effect of three factors. While two-way and three-way analyses of variance are beyond the scope of this basic treatment, it may be useful to consider a one-way analysis.

Example Z

Consider the sets of measurements of some property of a paint when manufactured by the four different techniques, A–D, given below. The data are arranged into groups, each group corresponding to a value (or level) of the factor in question, i.e. the processing conditions. The significance of this factor is tested by the value of the *F*-ratio of the variance due to the factor (that is, the mean square between groups) to the residual variation (that is, the mean square within groups).

The results for paints produced by the four processes are:

A	*B*	*C*	*D*
4	4	5	4
3	5	5	6
3	6	7	7
5	6	6	5

The means of the four groups are: A, 3·75; B, 5·25; C, 5·75; D, 5·5; the overall or grand mean is 5·063.

To perform the analysis, a series of computations must be undertaken and the following have to be evaluated:

Between-groups sum of squares
Within-groups sum of squares
Mean sum of squares, or variance estimate

The between-groups sum of squares is the sum of the squares of the deviations of the group means from the grand mean multiplied by the number of items in the group, i.e. 9·688 in this example. The between-groups degrees of freedom (for four groups) is 3, so that the between-groups variance estimate or mean square is 9·688÷3=3·229.

The within-groups sum of squares is the sum of the squares of the deviations of each item from its group mean, i.e. 13·25, while the within-groups degrees of freedom is (4–1)+(4–1)+(4–1)+(4–1)=12. Thus the within-groups variance estimate is 13·250÷12=1·104. It is now possible to calculate the variance ratio, *F*:

$$F = \frac{\text{between-group variance estimate}}{\text{within-group variance estimate}}$$

$$= \frac{3{\cdot}229}{1{\cdot}104} = 2{\cdot}925$$

This value of F does not exceed the theoretical F value, which at the two degrees of freedom (3, 12) is 3·49 at the 5 per cent and 5·95 at the 1 per cent level. Consequently, there is no statistically significant difference between the groups and one can conclude that there is no difference between the four processes. One could arrive at the same conclusion by performing a series of F and t tests, but obviously this is a far more tedious procedure.

There are other ways of performing an analysis of variance, particularly a 'shorthand' approach, but clearly a standard statistics text should be consulted for a detailed discussion of analysis of variance.

3.10.5 OTHER STATISTICAL TESTS

Various other tests are used in statistical analysis. Although they are important, space limitations prevent more than a brief mention here.

The chi-squared test

The chi-squared (χ^2) test is a method of determining whether the observed frequencies (or measured values) differ significantly from those expected under some assumed hypothesis. If the expected values are known, χ^2 is calculated:

$$\chi^2 = \Sigma \left(\frac{(O-E)^2}{E}\right)$$

where O is the observed value and E is the expected value, and all values are summed. Clearly, when there is a good fit between the observed and expected values, χ^2 will be small; the larger the value of the statistic, the poorer is the agreement. Tables of theoretical values for χ^2 at different degrees of freedom are available and the significance level of the result can be determined from these tables. The χ^2 test can also be used for measurements when the expected (E) values are not known and in fact fairly sophisticated analyses can be performed with this test.

Regression analysis

Often, in making measurements, it is useful to assess the degree of association or correlation between two sets of data. Typically, heights and weights, film permeability and pigment volume concentration of paints are examples of bivariate (two-variable) data populations. Plotting such data against each other will yield a graph with points that more or less lie along a straight (or possibly a curved) line, depending upon the closeness of correlation and the nature of the relationship between the data. It is possible to calculate the line of best fit to data, known as the 'least-squares line', as well as to calculate the correlation coefficient, r, for the data. Perfect fit (or correlation) is achieved when $r=1$; there is no correlation at $r=0$.

Furthermore, it is possible to calculate the statistical significance of r using Student's t test. The analysis outlined above is known as regression analysis and is useful in predicting the value of one parameter from another correlated parameter; it is obviously applicable to paint testing.

Non-parametric tests

The statistical tests covered so far have been concerned with parametric (that is, numerical) data. In many cases, however, it is not possible to assign exact or numerical values to the property that is under investigation—as, for example, in deciding on the weathering resistance or corrosion protection of coatings or aesthetic appeal of paints. In such cases, non-parametric data are being analysed and ranking tests have to be used. In fact, various non-parametric statistical tests are widely used for such analyses, the principal ones being the Wilcoxon and the Mann-Whitney tests.

These two tests are fairly simple to use in practice but are difficult to describe simply. The Wilcoxon matched pairs test uses identical pairs of specimens or the same specimens before and after testing. The property being measured is scored or given a value on some arbitrary scale (say, 0 to 5) and the change in score following testing is noted. The differences in scores are ranked or placed in order and then the ranks with the smaller total (i.e. either the negative or positive changes, whichever are the smaller) are summed to give a value T. This sum is compared with a tabulated theoretical value of T and the statistical significance of the changes or differences can be determined.

The Mann-Whitney U test differs in that paired differences are not required, and in fact it uses unrelated or independent samples, the results for the series being compared. Again the scores are ranked for the two series of n_1 and n_2 cases, but the ranking is for all samples tested. Then the ranks for n_1 samples are summed to give a sum value of R_1, and from this U is calculated:

$$U = n_1 n_2 + \frac{n_1 (n_1 + 1)}{2} - R_1$$

The derived value of U is then compared with the theoretical or critical value of U for the values n_1 and n_2. If the calculated U value is greater than the critical value for the selected significance level, then it can be concluded that there is no difference between the two series.

The above non-parametric tests, together with several others that are also used, are very useful in comparing data that are difficult to quantify. It is clearly impossible to discuss statistical tests in detail here but reference to the standard works listed in the bibliography will provide the reader with any further information that may be required. Various other items of information such as production figures, process yields, etc., can be analysed by statistical procedures to provide useful data, particularly for deciding upon production schedules, which are beyond the scope of this work.

3.11 Bibliography

Moroney, M. J., *Facts from Figures*, 3rd edn. Penguin Books: Harmondsworth, 1969.

Spiegel, M. R., *Theory and Problems of Statistics*. McGraw-Hill: New York, 1961.

Lamb, C. G., *Elements of Statistics*, 2nd edn. Longman: London, 1969.

Von Fraunhofer, J. A. and Murray, J. J., *Statistics in Medical, Dental and Biological Studies*. Tri-Med Books: London, 1976.

Siegel, S., *Non-parametric Statistics for the Behavioural Sciences*. McGraw-Hill: New York, 1965.

CHAPTER 4

Formulation of Coatings for Steelwork

4.1 Introduction

Carbon is the principal alloying element used for hardening and strengthening iron, and the resultant alloy, steel, is the most widely used metal in heavy and light engineering and the construction industry. The properties of carbon steels are largely dependent upon the proportion of alloying carbon added, which typically can vary from about 0·1 per cent to 1·2 per cent. Addition levels at the lower end of the range produce steels that are ductile and tough, strength being only a minor consideration; the higher levels produce very hard steels suitable for tools.

The carbon steel most often encountered by the paint technologist is mild steel, which contains from 0·15 to 0·25 per cent carbon, and this material is used where a combination of strength and machinability is required. This steel is largely used structurally for buildings, bridges and rolling-stock and for many light industrial applications.

Steel is produced by a number of methods, of which the traditional blast furnace–Bessemer process is used here as an example. The blast furnace, which is constructed of refractory bricks covered with steel plates, is charged with a mixture of iron oxide ore, limestone and coke. The coke is ignited at the bottom of the furnace by a blast of hot air, and the carbon monoxide so produced reduces the iron oxide to iron. The heat of reaction also decomposes the limestone into carbon dioxide and lime, and the latter combines with sand and other impurities in the ore to form a molten slag. The molten iron is tapped off preferentially at the bottom of the furnace. The obtained metal is known as pig iron or cast iron and contains up to about 4·5 per cent carbon.

The pig iron is run into the Bessemer converter and air is blown into the molten metal. This oxidises the carbon present in the mass, and the correct amount of carbon required for the steel is incorporated, generally by the use of an alloy of iron, manganese and carbon. The carbon present in the final steel is in the chemical form of iron carbide, Fe_3C.

After production, and during post-fabrication cooling, a complex layer of iron oxides known as mill-scale forms on the surface of steel, and this has to be removed prior to painting. Another characteristic of steel is its tendency to corrode or rust when exposed to the atmosphere, and again these air-formed oxide surface deposits have to be removed before painting. After removal of these bulk residues, generally by abrasion, it is normal practice in most situations to pretreat the surface of the steel prior to application of the paint system.

At its simplest, this pretreatment can involve solvent wiping, immersion in a liquid solvent or vapour degreasing to remove oils and greases, although complex phosphate pretreatments applied at high temperatures are frequently employed. As a full description of surface cleaning and pretreatment of steelwork is outside the scope of this work, the interested reader is referred to texts in the bibliography at the end of this chapter for further information.

4.2 Paint systems for steel

Once cleaned and pretreated, the steel is ready for coating. The type of paint system applied is clearly dependent upon the component itself, its anticipated life and service environment, as well as many other factors. The range of finishes applied to steelwork is vast and reflects the metal's diverse industrial applications. Space restricts the number of coating types that can be considered here, but a cross-section of the more commonly used systems is considered below.

A convenient method for subdividing the paint types used on steel is on the basis of where paint application is performed, i.e. whether in the factory (known as industrial finishing) or on-site; thereafter, a further subdivision can be the expected severity of the service environment. Such an approach has been adopted in this chapter.

Paints applied in the factory generally need to dry (or cure) rapidly, since this facilitates throughput of work. Coatings are therefore heated to effect cure, this process being known as stoving or baking. An alternative approach is to use catalysed paint systems which give rapid cure under ambient or low heat conditions. Speed again necessitates that the coating system is limited to the bare minimum of individual components consistent with obtaining the required finish and film thickness. Application of paints industrially is by spraying and, increasingly, by immersion techniques.

Several areas of industry make use of specialised techniques for applying paint and correspondingly highly sophisticated, purpose-designed coatings have to be used in such applications. Examples of such special techniques include powder coating, electrophoretic painting, electrostatic coating and radiation-cured coatings. Again, the complexity and highly specific uses of such coatings render their detailed description outside the scope of this text.

Paints applied under site conditions dry either by oxidative processes or solvent-loss (the so-called air-drying systems) or by catalyst-induced curing. These coatings are generally applied either by brushing or by spraying.

Exposure conditions

Exposure conditions can be subdivided into four categories, namely severe, moderate, mild and very mild. Clearly, coatings need to be selected and formulated to be able to withstand the most severe atmospheric conditions likely to be encountered in the location of the structure or article.

Severe conditions are typically total immersion in chemicals or salt water, coastal situations, areas of high industrial pollution and areas close to chemical processing plants.

Moderate conditions are found on outdoor exposure in areas of high rainfall and continuous high humidity in light industrial and urban conditions, rural areas and inland areas more than about 2 miles distant from industrial conurbations and coastal areas.

Mild conditions are experienced in interiors subject to condensation such as kitchens and bathrooms, as well as interiors where there is a source of pollution. Very mild conditions are found in warm dry interiors not subject to condensation or pollution.

In terms of coating performance on steel, the severity of the service environment dictates both the durability of the coating with respect to self-integrity and the ability of the system to reduce the corrosion rate of the metal substrate. Under severe conditions, corrosion of steel will be rapid and readily visible, will disrupt unsuitable coatings and may endanger the safety of the structure. In such conditions, where severe chemical attack by polluted atmospheres or liquid contact is expected, coatings of high performance such as chlorinated rubbers, epoxies and solvent-soluble vinyl resins are frequently used.

In moderate conditions, corrosion may become evident after one or two years' service and, although this will normally present no safety risk to the structure, it may be aesthetically unacceptable. In mild and very mild conditions, corrosion may only become apparent after a period of several years; however, it is normal practice when formulating paints for steel to adopt resistance to mild conditions as a minimum requirement.

The ultimate service environment therefore has major implications in paint formulation, for it is a basic requirement when painting steel to reduce the corrosion probability under the anticipated conditions to an acceptably low level for that situation.

Protection against corrosion

In practice, paints can reduce the corrosion rate of steel in a number of ways:

(a) through the formation of a relatively impervious surface layer on the metal, i.e. by barrier action;
(b) by the provision of a supply of inhibitor through leaching of inhibitive pigments from the paint coating;
(c) through the formation of an inhibitive agent by reaction of hydrolytic decomposition products from the polymer with constituents of the pigment to form a soap;
(d) by the inclusion in the paint coating of metallic powders which cathodically protect the metal substrate.

The simplest method is that of barrier action, where the imposition of a coating film between the steel and its environment prevents or markedly reduces access of the corrosion initiators, i.e. water, oxygen and aggressive soluble ions. In general, this approach is only of value where mild or very mild exposure conditions are encountered. However, much fabricated steel, such as office and domestic articles, is successfully protected using this technique. Such articles tend to need a relatively high cosmetic appeal. Furthermore, film characteristics other than durability and/or chemical resistance may be a prime requirement, for example, gloss- and colour-retention, and scrub- and abrasion-resistance.

In somewhat more aggressive situations, where there may be contact with liquid water or aqueous solutions for long periods, coating systems containing specially formulated corrosion-inhibiting primers are generally used. The corrosion inhibitor may function in any of the ways summarised in (b), (c) and (d), of which (b), the leaching of inhibitive components from the pigment, is one of the most appropriate techniques in such circumstances.

In particularly severe exposure conditions, enhanced protective ability is commonly provided in the formulation by means of components of the applied coating system other than just the primer. This is frequently accomplished by using topcoats having enhanced barrier action by means of very high film-build, by the incorporation of specific pigments and extenders, by the use of high-performance polymers, or by combinations of two or more of the above.

4.3 Industrial coatings

4.3.1 INDUSTRIAL COATINGS FOR MILD CONDITIONS

Finishes applied industrially and required to withstand only mild or very mild conditions are in general primarily required to provide decoration rather than protection, although clearly the latter is of importance.

Alkyd and alkyd–amino finishes

Typically, one-coat finishes are applied directly onto the cleaned and pretreated steel, and these are generally based on short or medium oil-length alkyd resins either used alone or in combination with amino resins such as melamine-formaldehyde and urea-formaldehyde. Such coatings are cured by stoving and the resultant films have good colour and gloss stability and high adhesion. The resistance of many such formulations to water, cleaning-agents, detergents, household chemicals, etc., is high and, in general, the films tend to be hard and resistant to abrasion and scrubbing.

Alkyd resins used with amino resins are typically based on non-drying oils such as coconut or castor oil, or semi-drying oils such as soya bean oil, at short or medium oil-lengths. The amino resin functions as a reactive intermediate to improve the hardness, chemical resistance and durability of the resultant film. These resin combinations, when used on steel, are normally stoved at high temperatures to effect film-formation. Acid-catalysed alkyd–aminos, which cure at ambient temperatures, are available although these are primarily of interest in wood-finishing formulations (see page 138).

In general, alkyd finishes containing melamine resins cure more rapidly than urea-containing types and, if necessary, they can also be stoved at temperatures up to 250°C. Urea–formaldehyde combinations, however, should not be cured at temperatures above 150°C. Alkyd–melamine resin combinations also have better initial gloss than alkyd–urea systems and they also exhibit greater colour stability on over-stoving. A further advantage is their superior chemical resistance and greater exterior durability.

Alkyd–melamine combinations do have certain disadvantages compared with alkyd–urea resins, particularly an inferior adhesion to steel. This therefore necessitates either the use of primers or replacement of some 20–30 per cent of the melamine content by urea resin, the latter modification generally not being detrimental to other film properties.

A wide variety of alkyd–amino resins ratios can be used, from about 1·0 : 1·0 to 1·0 : 0·1, and the resultant properties of the derived films are largely dependent upon the ratio used. Greater amino ratios produce harder films with better colour-retention on stoving although this is at the expense of flexibility and exterior durability. In general, films of alkyd–urea combinations are less hard than those of alkyd–melamine resins used at similar ratios.

Pigmentation. Pigmentation of these types of finish presents few problems. Where a full gloss is required, pigment volume concentrations of 16–20 per cent are typical, although lower levels are used with certain pigments, notably organics. Low film opacity, however, limits the use of many such pigments in one-coat finishes. Where lower gloss levels are required, low levels (less than about 1 per cent) of silica flatting agents can be used.

Table 4.1 Alkyd–Urea Stoving Finish, White

	Weight (%)
Titanium dioxide, rutile or anatase	22·2
Short-oil-length castor-oil-modified alkyd, 60% in xylene	27·8
Butylated urea-formaldehyde resin, 68% in *n*-butanol	16·6
Xylene	25·0
n-Butanol	8·4

The example of a white stoving finish given in Table 4.1 would be suitable for kitchen furniture, refrigerators and similar appliances.The pigment volume concentration of this formulation is about 17 per cent, and the ratio of alkyd to urea solids is 6:4. This type of formula could be applied direct to steel using conventional spray techniques, the viscosity being adjusted to a suitable level for the spray equipment with an appropriate balanced solvent 'thinner'. Stoving would typically be at 120°C for 30 minutes' duration.

Solvent. The solvent blend shown in Table 4.1 is characteristic of alkyd–amino systems. The alcohol solvent is required to maintain the stability of the amino resin and it also enables low paint viscosities with reasonably high solids content to be achieved. Xylene is necessary to solubilise short–medium-oil-length alkyds and butylated amino resins are also compatible with this solvent. Although other aromatic hydrocarbons would also fulfil this function, the use of xylene is almost universal in this application.

Manufacture. The manufacture of alkyd–amino coating systems needs some care in that the pigments must first be dispersed in the alkyd resin, the amino constituent being added only after dispersion has been achieved. Dispersing the pigment in the amino resin and then adding the alkyd would result in pigment flocculation and consequent reduced gloss level in the final film. This effect is due to destabilisation of the paste as a result of the more powerful pigment-wetting ability of the alkyd resin.

Special finishes

Speciality finishes, rather than the smooth types described above, are often applied to office, factory and domestic equipment and appliances. The most popular of these are the so-called wrinkle and hammer finishes.

Wrinkle finishes are prepared from short- or medium-oil-length alkyd resins modified with tung oil, a medium with a tendency to wrinkle rather than dry to a smooth finish if the drier combination is incorrect. To obtain a wrinke finish, therefore, manganese-metal-based driers, which aggravate

Table 4.2 Wrinkle Finish, Black

	Weight (%)
Carbon black	1·9
Diatomaceous silica	4·7
Short-oil-length tung–linseed-oil-modified alkyd, 60% in xylene	46·1
Xylene	45·6
6% Cobalt naphthenate	1·4
10% Manganese naphthenate	0·3

this tendency, are deliberately added to paints based on such media. Further control of the finish is achieved by the addition of low quantities of diatomaceous silica which modifies the 'size' of the wrinkle effect. Levels of about 0·1 per cent manganese metal drier on the resin solids and about 5 per cent silica on paint weight are common.

The finishes produced from these types of formulation are very tough and impact-resistant, these properties being largely attributable to the tung oil component of the binder. A formulation for a black wrinkle finish is shown in Table 4.2; this would be stove-cured for 1 hour at 120°C or for proportionately shorter times at higher temperatures. Application would be in one coat by spraying.

The use of wrinkle finishes, however, has been largely superseded by hammer-effect coatings; the term 'hammer finish' is used in allusion to the beaten-metal appearance of the final film.

Both air-drying and stoving variants of hammer finish are available and, in a similar manner to wrinkle finishes, they can be applied as one-coat finishes directly onto steel, generally by spraying. The hammered-metal effect is achieved by the addition of carefully controlled amounts of silicone fluid to the paint, and variation in the addition level controls the size of the 'hammer'. The metallic effect is further developed by the use of non-leafing aluminium paste. Pigmentation, if required, is limited to very low levels of organic pigments.

The formulation of a stoving-cured hammer finish based on an alkyd–urea resin combination is shown in Table 4.3, and this would be cured at 120°C for 20 minutes. Air-drying hammer finishes can be prepared from medium-oil-length drying-oil-modified alkyds, epoxy esters or vinyl-modified alkyds.

4.3.2 INDUSTRIAL COATINGS FOR MODERATE CONDITIONS

Where moderate exposure conditions are encountered on exterior exposure or even in certain interior situations, extra protection is required to be built into the coating system. Generally, primer coats are used prior to the finish

Table 4.3 Stoving Hammer Finish, Grey

	Weight (%)
Non-leafing aluminium paste	1·9
Short-oil-length dehydrated castor-oil-modified alkyd, 60% in xylene	68·1
Butylated urea-formaldhyde resin, 50% in xylene	22·5
Xylene	4·9
n-Butanol	1·9
Silicone fluid	0·2

Table 4.4 Single-pack Etch Primer

	Weight (%)
Lead chromate	9·0
Talc	1·4
Poly(vinyl butyral), low-viscosity grade	9·0
Isopropanol	60·6
Methyl ethyl ketone	13·9
85% Phosphoric acid	2·9
Water	2·9

Table 4.5 Two-pack Etch Primer

	Weight (%)
Base	
Zinc tetroxychromate	7·0
Talc	1·1
Poly(vinyl butyral), low-viscosity grade	7·2
Isopropanol	50·0
Toluol	14·7
Acid Component	
85% Phosphoric acid	3·6
Water	3·2
Isopropanol	13·2

and, again, the requirements of factory finishing necessitate that these are capable of being cured rapidly.

Etch primers

One of the most important types of primer used industrially (and also for site-applied coatings) is the etch primer, also variously known as a wash, pretreatment and self-etch primer. These are low-viscosity systems that can be applied to the substrate metal by spray, brush or dipping without further dilution. The film thicknesses produced have to be low, about 12–15 μm, in order to obtain optimum film properties. Consequently, a further primer coat is applied prior to finishing where long-term protection is required or when the exposure conditions are likely to be severe. In industrial coating, however, the finishing coat in many circumstances may be applied direct to the etch-primed surface with improved adhesion as well as greater corrosion-protection.

Etch primers are based on poly(vinyl) butyral resins in combination with specific chromate-inhibitive pigments and an acid, both single-pack and two-pack types being produced (Tables 4.4 and 4.5). Under ambient conditions the drying time of etch primers is quite short; in industrial use accelerated drying can be achieved by the use of moderate-temperature stoving.

Optimum results are achieved with etch primer formulations if the ratio of active pigment–resin solids–acid is maintained fairly close to 1 : 1 : 0·5. In the examples given in Tables 4.4 and 4.5 the solution viscosity characteristics of the resin necessitate the use of the specified relatively low solids contents. The formulation of etch primers is somewhat critical and, although fairly major adjustments can be made to solvent content, for example to modify the flash-point, the level of the other ingredients should not be altered markedly.

Pigmentation. The active pigment is always a low solubility chromate. Zinc tetroxychromate is used in the two-pack materials; in single-pack formulations lead chromate is preferred since it increases the stability of the composition. Once mixed, the two-pack etch primers must be used within 8 to 12 hours. The single-pack types are normally stable for periods of 12–18 months.

Binders. Although poly(vinyl butyral) resins are always used in etch primers, modification with alcohol-soluble phenolic resin at levels of up to one-third substitution of the total resin is sometimes practised in order to improve the water-resistance of the film. Low-viscosity grades of poly(vinyl butyral) are used in order to maximise the solids content of the resultant primers.

Solvents. Alcohol solvents are required for solubilisation of poly(vinyl butyral) resins although water has to be added in order to achieve low

Table 4.6 Stoving Red Oxide Primer

	Weight (%)
Red oxide of iron	29·0
Zinc chromate	8·2
Talc	2·4
Whiting	5·7
Medium-oil-length styrene modified alkyd, 50% in xylene	31·4
Xylene	23·2

viscosity solutions. Besides improving the solvation, water is essential to confer stability on the formulation. With a water content of less than about 2·5 per cent, gellation will occur within minutes of mixing the acid component: too high a level, greater than about 6·5 per cent, causes flocculation of the pigment with reduced protection. The organic solvents used are generally alcohol blends, normally ethanol, *n*-butanol or isopropanol, in conjunction with aromatic hydrocarbons, esters or ketones.

Other primers for industrial application
Where more conventional primers are required for industrial applications, formulations similar to those used for the site painting of steelwork are employed with suitable changes to enable the film to be stove-cured. One such formulation is shown in Table 4.6 for a red oxide–zinc chromate industrial primer based on a vinyl-modified alkyd resin. This particular example could be spray- or dip-applied and cured at 120°C for 30 minutes. The pigment volume concentration is about 42 per cent.

Binders. With industrial primers, provided that the adhesion of the finishing coats is not impaired by the film hardness developed during stoving, there is little restriction on the resin base that can be used. For example, primers can be based on suitable modified or unmodified alkyds, epoxy esters, alkyd–aminos, acrylic–aminos, epoxy–aminos, or on epoxy–phenolics for more demanding situations.

Pigmentation. Primer pigmentation can also be varied within wide limits although formulations based on the corrosion-inhibitive pigments commonly used in paints for on-site application are seldom required in industrially cured coatings. Protection is imparted mainly by the chemical resistance and low permeability of the resin system used. Zinc chromate and zinc phosphate are the most frequently used inhibitive pigments in stoving primers.

Primers for many industrially applied coatings are in fact frequently based on combinations of non-inhibiting inert pigments and extenders. A range of

Table 4.7 Alkyd–Melamine Exterior Finish, White

	Weight (%)
Titanium dioxide, rutile	22·5
Medium-oil-length drying-oil alkyd, 50% in xylene	47·5
Butylated melamine–formaldehyde resin, 60% in *n*-butanol	9·5
Xylene	13·0
n-Butanol	7·3
6% Cobalt naphthenate	0·2

extenders is in common use in industrial primers, namely hydrated aluminium silicate (china clay), calcium magnesium carbonate (dolomite), calcium carbonate (whiting), barium sulphate (barytes) and lamellar hydrous magnesium silicates (talcs). Pigments for primers are typically inorganic in composition, and can be added at levels of up to about 20 per cent pigment volume concentration. Extenders are generally added to bring system P.V.C.s into the 50–60 per cent range, and typically talc or china clay is used in admixture with dolomite, whiting or barytes at a volume ratio of about 1 : 3.

Thickening agents. Where required, thickening agents such as amine-treated clays can be used to control the system rheology, and levels of about 0·1 per cent on paint weight are generally adequate for this.

Alkyd–amino finishes

The formulation of stoving industrial finishes to withstand moderately severe exposure conditions follows similar lines to the alkyd–amino finishes described in the previous section. Alkyd melamine combinations can be used successfully in this situation, applied over a suitably primed substrate. In general, relatively low levels of melamine resin are used in conjunction with medium-to-long oil-length drying-oil-modified alkyds, and this produces films with good exterior durability.

The formulation shown in Table 4.7 is typical, with a pigment volume concentration of about 16 per cent and a ratio of alkyd to melamine resin solids of 8 : 2. Stoving conditions of 30–40 minutes at 120°C would be required.

The finishing of automobile bodies constitutes an important application where both good exterior durability and corrosion protection are required in the coating system. The traditional finishes for bodywork were based on nitrocellulose lacquers (see page 141 for formulation details of similar coatings) but these have largely been superseded by more sophisticated coating systems. Both alkyd–melamine and, increasingly, thermosetting acrylic-resin-based paints are now used for automobile finishes, the formulation of the former types being similar to the example given in Table 4.7.

Thermosetting acrylic resin finishes

Thermosetting acrylic resin systems are used in conjunction with amino resins, most frequently melamine-formaldehyde types. The acrylic resins used are typically prepared by copolymerisation of esters of acrylic and methacrylic acid monomers. The ultimate hardness of the resultant acrylic resin is dependent upon the ratios used: methacrylic ester monomers impart greater hardness compared with acrylic esters. Where melamine resins are to be used with the thermosetting acrylics, further modification of the acrylic copolymer by chemical addition of monomers with pendant hydroxyl and carboxyl groups is common. The hydroxyl site in the resultant polymer forms a centre for reaction and cross-linking with the amino resin, while the residual acidity due to the carboxyl groups assists the curing process.

Paints prepared from these combinations are typically cured by stoving at temperatures of about 120°C. However, some modification enables the bake–sand–bake process to be performed. By this technique, minor repairs and blemishes in the paintwork can be sanded out after the coating has received a preliminary low-temperature curing treatment of about 20 minutes at 80°–100°C. After sanding, a final stoving period of about 30 minutes at 120°C causes the film to reflow and mask the damaged areas without any loss of gloss. The full hardness of the film is developed with this final period of stoving.

Table 4.8 Reflow Thermosetting Acrylic Finish, White

	Weight (%)
Titanium dioxide, rutile	22·2
Hydroxy-containing acrylic resin, 60% in xylene/*n*-butanol	32·6
Butylated melamine-formaldehyde resin, 70% in *n*-butanol	11·8
Xylene	20·3
2-Ethoxy acetate	7·7
n-Butanol	5·4

The formulation shown in Table 4.8 is for a white reflow thermosetting acrylic finish with which this technique can be used. The pigment volume concentration of about 19 per cent is typical for automobile topcoat formulations, the useful working range being in the region of 8–20 per cent with lower levels possibly being adopted for certain organic pigments. The ratio of acrylic to melamine resin solids of 7 : 3 can also be considered typical for this application.

Alkyd–amino and amino–acrylic automobile finishes are typically applied by spray in two or three coats over a primer–surfacer and corrosion-inhibitive primer.

4.3.3 INDUSTRIAL COATINGS FOR SEVERE CONDITIONS

Many types of paint formulation suitable for use in severe exposure conditions can be prepared. For industrial applications, stoving two-pack epoxy and polyurethanes as well as solvent-soluble and dispersion vinyl resin systems are used. However, in many cases structural steel and large fabricated sections are coated under controlled industrial conditions, the paints used being the air-drying types described in the following section. These coatings are typically derived from chlorinated rubber or vinyl systems, or catalysed systems based on polyurethane or epoxy, and a wide range of inhibitive and protective pigments and extenders are used. Speed of drying is not normally critical to these systems, since the close control that can be exercised on surface pretreatment and coating application is of greater importance.

Epoxy–phenolic systems

Industrially applied stoving-cured systems for exposure to severe conditions are typified by epoxy–phenolic systems which are used to coat the interiors of food-cans and chemical containers and other surfaces where excellent chemical resistance is required. Poly(vinyl chloride) organosols and plastisols, which are capable of providing high chemical resistance as well as long term exterior durability, are increasingly used in such applications.

The epoxy–phenolic systems are used where maximum chemical resistance is required. The epoxy resins used are high-molecular-weight solid types (where n=8 or 12), and the most suitable phenolic resins are the resol type with a low degree of condensation. The ratio of phenolic to epoxy resin is important: optimum performance with respect to chemical resistance and flexibility is obtained with ratios between 30 : 70 and 20 : 80.

During manufacture of these types of coatings, refluxing is required to achieve compatibility and to develop the film-forming properties of the system. This pre-reaction is carried out by heating the two resins together in solution for periods of 1–3 hours at temperatures of the order of 110°–150°C. If refluxing is not possible, film-formation can be improved by the use of flow promoters and additives to decrease cissing and pin-holing. Additions of cyclohexanol at 2–10 per cent of the solvent content and poly(vinyl butyral) at about 1 per cent on total weight are beneficial in this situation.

A formulation for an epoxy–phenolic clear finish is shown in Table 4.9. Typical stoving schedules for this type of system vary between ½ hour and 1 hour at 175°–200°C, although low levels of addition of acid accelerators, such as salicylic or phosphoric acid, can be used to allow a lower stoving temperature or effect very fast cure rates. When the finish is applied by a multi-coat process, it is normal to undercure the coats until the final coat is applied, at which stage the full stoving schedule is used. This improves the intercoat adhesion of the system.

Table 4.9 Epoxy–Phenolic Clear Finish

	Weight (%)
Diphenylol propane-based solid epoxy resin (epoxy equivalent 2400–3400)	30·4
Resol phenol-formaldehyde resin, melting-point 80°–86°C	13·2
2-Ethoxy acetate	17·4
Xylene	17·4
Diacetone alcohol	10·8
Toluol	10·8

Pigmentation. Pigmentation of this type of coating is rather restricted on account of the dark colour imparted by the resin components. When required, however, primers and topcoats can be readily prepared. The use of inhibitive pigments in the primer coats is not normally necessary, since the chemical inertness and very low permeability of the polymer component provides sufficient protection for most purposes.

Where better colour is required in pigmented systems, combinations of amino and epoxy resins are used, although this is at the expense of some chemical resistance in the film. Similar ratios to those described above are used and, with urea-formaldehyde resins, no precondensation process is necessary in order to develop compatibility. Melamine-formaldehyde resins are used with epoxy resins, but precondensation is essential with these systems. Stoving times of the order of 30 minutes at 150°C are typical with these coatings.

Dispersion coatings

Dispersion coatings such as organosols and plastisols are pigmented systems based mainly on poly(vinyl chloride) resin, although poly(vinyl chloride) copolymers can also be used. Application thicknesses can be very high with these systems, so that one-coat finishes can be achieved. Film cure is effected by a high-temperature fusion process.

Both organosols and plastisols can be considered as dispersions of powdered resin in a continuous liquid phase, and the major advantage of these systems is that they permit the use of polymers of much higher molecular weight than would be possible with solution vinyl resins. This produces films with superior mechanical and chemical resistance characteristics. A plastisol consists of a resin and a plasticiser, whereas an organosol consists of a resin, a plasticiser and a volatile liquid. Fairly low pigment levels can be used in dispersion coatings since adequate opacity is ensured by the high film-builds obtained.

Table 4.10 Dispersion Coatings, White

	Organosol (weight %)	**Plastisol** (weight %)
Titanium dioxide, rutile	10·0	10·0
Poly(vinyl chloride) powder	50·0	50·0
Plasticiser	19·0	39·0
Dispersant	10·0	—
Diluent	10·0	—
Stabiliser	1·0	1·0

The generalised formulations of a white organosol and a plastisol coating are shown in Table 4.10.

The resin component of dispersion coatings is used in powdered form, but because of aggregation effects it has to be reduced to its ultimate particle size during manufacture of the paint. This is normally accomplished using ball-milling techniques. The particle size required is in the region of 0·05–3 μm, the lower end of the range being used for the organosols to reduce settling, and the larger particle sizes for the plastisols.

A prime requirement for the liquid components is that at room temperature, and during storage, they should not have any swelling effect on the polymer powder. However, at the fusion temperature obtained during curing of the applied coating, they should solvate the dispersed particles and facilitate the coalescence of the polymer to form a coherent film.

Plasticisers. A wide range of plasticiser compounds is used in dispersion coatings and, since the final film will, apart from pigment and stabiliser, consist of resin and plasticiser, the amount and type used will exert a marked effect on the physical characteristics and chemical resistance of the coating. Materials such as dioctyl phthalate, dioctyl adipate and trichloro-phosphate are used, and the plasticiser–resin ratio can be varied between 1 : 2 and 1 : 1. The choice of plasticiser is especially critical in plastisol coatings where it must impart the required rheological characteristics as well as satisfy the necessary film requirements. Combinations of plasticisers are frequently used in plastisols.

Solvents and diluents. The volatile component in an organosol is based on both polar and non-polar materials, and the convention in dispersion-coating technology is to refer to the former materials as dispersants (or solvents), and the latter as diluents. The function of the polar dispersants is to wet the polymer particles and form a stabilising sheath around them, although it should not act as a solvent at room temperature. At the fusion temperature, however, its solvent effect should be pronounced so that it softens or even dissolves the polymer before it evaporates from the film. The selection of the dispersant is critical in determining the application characteristics, viscosity and stability. A range of materials can be used, including slow-evaporating esters, glycol ethers and ketones, particularly di-isobutyl ketone.

The diluents are aromatic or aliphatic hydrocarbons, or blends thereof, and they control the viscosity of the dispersion as well as modifying the action of the dispersant. The ratio of diluent to dispersant is important in maintaining the storage stability of the dispersion.

Stabilisers. The stabiliser is added to reduce the thermal degradation of the poly(vinyl chloride) polymer that would otherwise occur at the fusion temperature. Lead compounds, cadmium stearates, organotin compounds and oxidised oils are used for this purpose, often in combination. Despite the presence of stabilising additives, overstoving of dispersion coatings is to be avoided.

Pigmentation. Pigmentation of dispersion coatings is normally accomplished by means of inorganic pigments, but iron- and zinc-containing pigments are not used since these promote thermal decomposition of the polymer. High oil-absorption pigments must also be avoided since these absorb plasticiser, with a consequent requirement to increase the fusion conditions. The total pigmentation can be kept relatively low (Table 4.10), since the high application weight typically applied will promote opacity.

Application. Dispersion coatings can be applied by several techniques, spraying, roller-coating or dipping being most common. Dry-film thicknesses of the order of 75–250 μm are obtained, the higher thicknesses being achieved by the plastisols. Stoving schedules of about 15 minutes at 175°C are used, but the precise fusion temperature is dependent upon the formulation ingredients.

Because of the inert nature of poly(vinyl chloride), adhesion to steel is not of a high order and therefore it is usual to use primers to obtain adequate bonding. Stoving primers based on phenolic–epoxy combinations can be used successfully.

4.4 Site-applied coatings

4.4.1 SITE-APPLIED COATINGS FOR MILD CONDITIONS

Relatively little steelwork is specifically coated on-site to withstand very mild corrosive conditions. In general, the minimum formulation requirement for coatings applied under site conditions is for mild exposure severity.

Coating systems for mild conditions might typically include a primer which would be based on red oxide of iron, ideally in combination with the corrosion-inhibitive pigment, zinc chromate. This primer would invariably be overcoated with an undercoat followed by a finish combination formulated with an air-drying alkyd resin system. This type of coating system is widely used on steelwork of all descriptions and, with only slight formulation modifications, the performance of the components can be improved so that the system is adequate for moderate exposure conditions.

Primers

The red iron oxide–zinc chromate primers are generally based on alkyd resins or oleoresinous varnish binders although, increasingly, the use of emulsion binders, either pure acrylic or styrene–acrylic types, is becoming widespread in primer formulation.

Table 4.11 Air-drying Red Oxide–Zinc Chromate Primer

	Weight (%)
Red oxide of iron	50·4
Zinc chromate	10·1
Whiting	5·0
Long-oil-length linseed–tung-oil-modified phenolic varnish, 60% in white spirit	25·2
White spirit	8·1
24% Lead naphthenate	0·3
6% Cobalt naphthenate	0·1
Clay gellant	0·5
n-Butanol	0·3

A formulation for a brushing red oxide–zinc chromate primer based on oleoresinous varnish is shown in Table 4.11. This primer has a pigment volume concentration of about 50 per cent and a ratio of red oxide to zinc chromate of 5·0:1·0. Pigment volume concentrations used with such primers range approximately from 35 to 55 per cent, the lower end of the range being more suited for the higher zinc-chromate-content types, and the higher end for the low zinc chromate-content systems.

The ratio of red oxide to zinc chromate can be varied over a wide range, from formulations based entirely on red iron oxide to those containing only the zinc chromate pigment. The former, based entirely on a non-corrosion-inhibiting pigment, can be regarded as being suited for use only in the mildest of conditions, whereas those based entirely on zinc chromate provide a much higher level of protection and are suited for more severe exposure conditions. Primers based solely on zinc chromate are also preferred for use on aluminium metal, generally in conjunction with a self-etch primer. However, in primers for use on mild steel in mild exposure conditions, ratios of red oxide–zinc chromate of about 5:1 by weight are probably typical.

There is little restriction on the type of resin used in these primers, although a low level of free acidity is essential to avoid reaction with the basic zinc chromate, which would result in gellation of the liquid paint. Alkyd binders are most commonly used, and these should be of long oil-length in order to maximise durability. They are typically based on drying oils such as linseed or soya. Amine-treated clay gellants are often used to control the rheology of these primers and the level shown in Table 4.11 is adequate for many brushing-grade primers of this type.

Alkyd finishes

The undercoat and topcoat formulations applied over such primers are invariably based on alkyd binders and pigmented with colouring and opacifying materials of a non-inhibitive nature.

The oil-modified alkyds are one of the most widely used polymer types in paint formulation. These binders are highly versatile, being incorporated in many types of paint system for both decorative and protective functions. In their simplest form, alkyd resins can be considered as being the reaction product of a poly-carboxylic acid, such as phthalic anhydride, and polyhydric alcohol, such as glycerol. This is then modified during manufacture with a vegetable oil, or the fatty acid component of a vegetable oil, to yield the final alkyd polymer (see Chapter 1). The type of oil modification, whether drying, semi-drying or non-drying, as well as its addition rate, largely determines the properties and applications of these oil-modified alkyds.

The drying or semi-drying long-oil-length alkyds used in these decorative finishing coats impart high durability with rapid drying, a pale colour, and good colour and gloss stability. However, resistance to chemicals, water and heat is not a characteristic of such oil-modified systems. In more severe exposure regimes, where these latter characteristics are of greater importance, alkyds chemically modified with polyurethane, vinyl, epoxy or silicone resins tend to be used in coatings.

The formulation of a white oil-modified alkyd topcoat with a high gloss finish is shown in Table 4.12, and this type of paint can be used on many

Table 4.12 Alkyd Gloss Finish, White

	Weight (%)
Titanium dioxide, rutile	27·0
Long-oil-length soya-oil-modified alkyd, 70% in white spirit	60·0
White spirit	10·8
24% Lead naphthenate	0·86
6% Cobalt naphthenate	0·34
4% Calcium naphthenate	1·0

substrates other than mild steel. In a similar fashion, the compatible white undercoat, Table 4.13, is also widely used. Primers, however, are of necessity formulated for a particular substrate or combination of substrates.

Pigmentation. Full gloss finishes are favoured for exterior use since they tend to have an acceptable cosmetic appeal in most situations as well as high durability. To accommodate this requirement in topcoat formulations, it is generally necessary to omit extender materials completely and to work at pigment volume concentrations only slightly above the minimum required to produce adequate opacity. For example, the formulation given in Table 4.12 has a P.V.C. of about 14 per cent and this is normal for general-purpose white alkyd finishes. Different coloured pigments would possibly require different addition rates, although P.V.C. levels of about 15–20 per cent are rarely exceeded.

Solvents. White spirit is the solvent most frequently used in this type of alkyd-based formulation, since its evaporation rate is suitable for these brush-applied air-drying paints. Other aliphatic hydrocarbons are sometimes used.

Table 4.13 Alkyd Undercoat, White

	Weight (%)
Titanium dioxide, rutile	21·8
Barytes	29·5
Mica	5·5
Surface-coated calcium carbonate	5·5
Long-oil-length soya-oil-modified alkyd, 75% in white spirit	22·1
White spirit	15·1
24% Lead naphthenate	0·33
6% Cobalt naphthenate	0·17

Driers. The other main additives in alkyd finishing paints (and alkyd paints in general) are driers, and these are required to modify the oxidative drying of the resin to within acceptable limits, typically touch-dry in 4 hours and hard-dry in 16 hours. The drying acceleration process should also be accomplished without adverse side-effects on colour stability or subsequent film durability and, in general, metallic salts of naphthenic acid, the so-called naphthenate driers, are perhaps the most widely used types and offer a good compromise of properties.

With alkyd paints, cobalt metal driers used alone can produce satisfactory results. However, the rather high levels required, about 0·1 per cent metal on solid resin (and even up to 0·4 per cent in the case of paints containing pigments such as carbon black which can absorb driers), can cause the finish to acquire a disfiguring surface bloom under cool, humid conditions. Combinations of lead and cobalt metal driers in the ratio 10:1 are rather more effective and these permit lower cobalt levels. The addition of lead, however, is not always sufficient to suppress blooming entirely; further improvements can be achieved using calcium-based materials, although these have no effect on the drying process itself. Additions of calcium 'driers' are of course necessary only in topcoat formulations.

The overall drier addition rate has to be controlled quite closely since adverse effects on, for example, durability and gloss stability can be experienced if addition rates are excessive. Typically, a long-oil-length alkyd requires 0·25–1·0 per cent lead and 0·025–0·1 per cent cobalt, calculated as metal on resin solids; shorter oil-lengths require proportionately less and, conversely, oil paints need somewhat higher addition rates.

In many instances, lead metal has to be omitted from paint formulations for health and safety reasons and, accordingly, a lead-free drier system may be required. Zirconium-based driers are commonly used for this purpose as replacements for the lead component of the drier system, although addition levels can be up to 50 per cent higher. Most importantly, however, they are highly specific, their drying ability being dependent upon the composition of the alkyd resin.

Undercoats

Pigmentation. The undercoats for alkyd gloss finishes are formulated using similar principles to the above although much higher pigmentation levels are adopted (Table 4.13), since this component of the paint system is required to obliterate the colour of the previous coats as well as to provide a suitable ground colour for the finishing coat. The latter is of especial importance where low-opacity pigments have to be used in the topcoat. Furthermore, the undercoat needs to provide a mechanical key between the primer and the finishing coats as well as to increase the build of the paint system; here again, higher pigment contents are beneficial.

Pigment volume concentrations for air-drying alkyd undercoats are typically in the region of 40–60 per cent although there is no doubt that the higher levels can decrease the performance of the system. In particular, the ability of the film to accommodate substrate movement is limited, especially after a period of weathering, and this effect, while it may not be of prime importance on steel, has major repercussions when this type of alkyd system is applied to timber.

The pigment volume concentration of the undercoat example given in Table 4.13 is about 48 per cent. Higher P.V.C.s can be obtained by reducing the resin content and increasing the extender loading while maintaining the content of true pigment. This approach tends to result in a similar level of opacity in the resultant film at a somewhat lower cost.

Control of settling. The high pigment-extender content required in undercoats necessitates some control of the settling rates during storage. Low levels of clay gellants are sometimes used for this purpose, although care has to be exercised to ensure that the flow properties are not adversely affected. An alternative approach is to include surface-coated calcium carbonate at about 5–10 per cent of the total pigment content. This reduces the packing tendencies of the pigment and extender particles during settling and facilitates their subsequent reincorporation by stirring.

Aluminium paint

The alkyd-resin-based paint systems are probably the most widely used types on mild steel structures that are required to withstand mild and moderately severe exposure conditions. The pigmented appearance, however, may not be appropriate for every application and the use of metallic finishes, particularly aluminium paint, is a fairly common alternative.

These metallic paints, as their name suggests, derive their appearance from the inclusion of aluminium powder into a binder which must have a low free acidity in order to eliminate adverse reactions during storage. The protection afforded by such paints results from the use of leafing-grade aluminium which confers low permeability to water upon the paint system.

To maximise the leafing characteristics of the aluminium platelets, and hence the protective ability of the film, certain formulation requirements have to be satisfied. The resin component, for example, needs a lower viscosity than would be considered suitable for many applications (about 0·1 N s m^{-2}, 1·0 poise). It also requires a high surface tension, which can be aided by the use of a proportion of an aromatic hydrocarbon solvent. The wetting properties should be low and, since the two properties are related, this necessitates the use of medium–long-oil alkyds and varnishes of low acid value, or neutral resins such as coumarone. When oxidative-cured binders are used, the inclusion of lead driers is avoided since these suppress the

Table 4.14 Aluminium Paint

	Weight (%)
Leafing-grade aluminium paste, 70% powder in white spirit	18·8
Coumarone resin, 50% in white spirit	25·8
Linseed stand oil, 100 poise	26·9
White spirit	22·5
Xylene	5·4
6% Cobalt naphthenate	0·6

leafing action of the aluminium. Resin contents are usually kept high and this also improves both the appearance and the performance of the aluminium paint. Addition levels of 10–20 per cent of aluminium metal are typical.

A typical air-drying aluminium paint formulation is shown in Table 4.14, and this would be brush- or spray-applied (without dilution) in two coats onto the cleaned steel. Again, this type of finish is suited for use on mild steel to be used in mild and moderately severe exposure conditions. Alternatively, and again in a similar manner to the alkyd-based paints previously described, a similar formulation can be included in high-durability paints for steelwork. For example, aluminium powder can be incorporated in pigmented anti-corrosive paints to reduce the water-permeability characteristics of the system.

4.4.2 SITE-APPLIED COATINGS FOR MODERATE CONDITIONS

Modified alkyd resin paints

As indicated in the previous section, the use of conventional drying-oil–alkyd paint systems is acceptable for both moderate and mild conditions of exposure. Frequently, however, modified alkyd resins can be used in undercoat and topcoat formulations, often in conjunction with more protective primer formulations. This modification may be copolymerisation, typically the addition of polymer precursors such as styrene, vinyl toluene, silicone or isocyanate during alkyd manufacture, or by physical blending during paint manufacture with modifying resins such as chlorinated rubber. The films produced from the latter modification are suitable for severe exposure conditions as maintenance paints, and for convenience they will be discussed in the next section.

The chemically modified alkyds do possess certain advantages which render their use desirable in specific circumstances. The use of vinylated alkyds, either styrene or vinyl toluene types, produces films that are quick-

Table 4.15 Red Lead Primer

	Weight (%)
Non-setting red lead	79·4
Asbestine	4·0
White spirit	2·0
Acid-refined linseed oil	13·9
6% Cobalt naphthenate	0·7

drying and harder compared with the conventional oil-modified alkyds. Silicone modification upgrades the water and heat resistance of the film and isocyanate-modified alkyds are quick drying and tough as well as having superior resistance to solvents and chemicals compared with conventional types.

In general, however, these benefits are countered by certain disadvantages. For example, most modified alkyds have a tendency to chalk (erode) rather more rapidly than unmodified types, and the colour retention is less good. Paints based on these modified alkyds are formulated in a broadly similar manner to the conventional drying-oil–alkyd based paints described earlier.

Lead primers

The use of primers other than the red iron oxide–zinc chromate types previously described is now limited by the fact that lead was included in the pigmentation of many of the alternatives, either as elemental lead or a lead salt. The increasing awareness of the health risk attached to the use of these lead-containing primers has rendered their future use problematical and, indeed, in certain situations they have been fully replaced by alternative primer types. However, there is no doubt that the traditional red lead primers, especially when based on linseed oil, must still rank as the finest anti-corrosive primers that can be formulated for on-site application to mild steel, particularly where the surface preparation cannot be of the highest standard.

The formulation of a red lead–linseed oil primer is given in Table 4.15, and this conforms to British Standard BS 2523: 1966, Type A. This standard specifies three primer formulations, Types A, B and C, which differ in their red lead content. Type A contains about 80 per cent red lead by weight and Type B about 60 per cent by weight, and these are recommended for steelwork to be left in primer for very long periods before topcoat application. The Type C formulation is a blend of red and white leads, and the presence of the latter pigment increases the rate of drying to the extent that films of this primer can be overcoated after about 24 hours.

The pigment volume concentration of the formulation shown in Table 4.15 is about 38 per cent, and the asbestine additive functions as a suspending agent to reduce the rate and degree of compaction of the pigment. This primer would be brush applied and could take several days to dry, depending on the weather conditions. The speed of drying of these lead primers can be improved by the use of alternative binder systems, and this is also of importance where a compatible coating system is required. Binders such as alkyds and oleoresinous varnishes can be used; chlorinated rubber or epoxy resin systems are adopted for severe exposure conditions. A minimum pigment volume concentration of 30 per cent is generally considered desirable in paints formulated on these other types of resin.

Lead-based primers containing metallic lead or calcium plumbate are also known, the latter being used on galvanised steel. Metallic lead primers can be formulated on a wide range of binders and, as with red lead, there are no major formulation problems. With metallic lead primers, optimum performance is achieved at a pigment volume concentration of about 40 per cent, although lower levels of about 30 per cent P.V.C. have been used in chlorinated rubber-based primers, while higher levels of about 50 per cent may be incorporated in the oleoresinous varnish type of binder.

Pigment volume concentrations of about 35 per cent in oil or alkyd media are favoured for calcium plumbate and, as with other lead-containing primers, the level of the corrosion-inhibitive pigment itself constitutes a substantial proportion of the total pigmentation.

Less toxic types of primer

Because of the high toxicity risk of these lead primers, alternatives are becoming widely used. Pigments such as red iron oxide–zinc chromate mixtures, or those based entirely on zinc chromate, can be used, often in conjunction with an etch primer (see page 99), in moderately severe exposure conditions.

Increasingly, completely non-toxic metal primers based on zinc phosphate pigment are used and these do have certain advantages. Unlike the other corrosion-inhibiting pigments, zinc phosphate is white in colour and has a low opacity. Consequently, this pigment permits the formulation of light-coloured or even white primers. Because of its chemical inertness, zinc phosphate can be used with a wider variety of binder types than the other anti-corrosive pigments although, as a consequence, zinc phosphate primers are perhaps less effective than other types. Pigment volume concentrations of zinc phosphate primers are generally maintained in the region of 30–40 per cent, although higher levels can be used. The zinc phosphate itself generally comprises 50 per cent or more by weight of the total pigmentation, the remainder being colouring and opacifying pigments and low levels of extender added to aid suspension or to modify the film strength or permeability characteristics.

4.4.3 SITE-APPLIED COATINGS FOR SEVERE CONDITIONS

A wide spectrum of coating systems is used to protect steelwork subjected to severe conditions of exposure. Typically, the binders used are amongst the most sophisticated available, and all can be characterised as having very high (though specific) chemical resistance, as well as proven long-term durability. Chlorinated rubbers, vinyl copolymers, polyurethanes, epoxies and inorganic silicates are foremost among those used, and it is usual to match the formulation of the various components fairly closely in coating systems for demanding exposure conditions.

A whole range of pigment types is used in these formulations, although in certain exposure situations it is necessary to ensure that the pigmentation adopted is itself stable. In many areas of use, protection will be formulated into all components of the system. While the use of inhibitive pigments is restricted to primer coats, flake pigments and extenders such as micaceous iron oxide, aluminium, stainless steel, mica and certain talcs are often incorporated into finishing coats.

Both conventional paint systems, i.e. a primer, undercoat and topcoat, as well as high-film-build technology are encountered in formulations for severe conditions. Some systems are required both to have good durability and to be decorative, whereas others can be of lower cosmetic value. This consideration clearly has major implications in the selection of materials for the respective formulations.

Chlorinated rubber-based systems

One of the most widely used binders for paints to withstand severe exposure conditions on steelwork is chlorinated rubber. A useful attribute of this binder is that it dries by solvent loss and, consequently, the derived coatings can be used in situations where low ambient temperatures during paint application would otherwise adversely affect the curing process of chemically-cured paints, such as those based on epoxies or polyurethanes. Paints based on chlorinated rubber are used in the formulation of both low- and high-build coatings for steelwork.

Alkyd–chlorinated rubber blends. Interesting variants of low-build chlorinated rubber paints can be obtained by blending this polymer with alkyd resins and, despite the presence of the latter in the binder, the performance of these systems as general-purpose maintenance paints in chemical and industrial plant and factories, etc., can be very good. A formulation of a white gloss finish maintenance paint based on this combination of resins is shown in Table 4.16, and general formula considerations are discussed on page 154. This paint, of about 14 per cent P.V.C. and a ratio of chlorinated rubber to alkyd of 1:3, could be brush-applied over a compatible undercoat–primer as a two- or three-coat system, the primer containing an inhibitive pigment. The white spirit in this

Table 4.16 Alkyd–Chlorinated Rubber Topcoat, White

	Weight (%)
Titanium dioxide, rutile	18·4
Long-oil-length soya-oil-modified alkyd, 70% in white spirit	36·6
Chlorinated rubber, low-viscosity grade	8·5
Xylene	26·9
White spirit	9·0
24% Lead naphthenate	0·4
6% Cobalt naphthenate	0·2

formulation reduces the risk of film rivelling during application of the system although, conversely, the aromatic solvent improves the intercoat adhesion when applied to weathered, hardened films. Where higher chemical resistance is required in the system, the amount of alkyd can be reduced to give a ratio of chlorinated rubber to alkyd as low a 1:1. In such cases, however, extra plasticiser, typically in the form of a chlorinated paraffin, is desirable, particularly if the chlorinated rubber exceeds 30 per cent of the binder.

Chlorinated rubber–chlorinated paraffin resin–plasticiser combinations often form the bases of paint systems where high performance is required, and both low-build and high-build formulations are widely used.

Chlorinated rubber-based primers can be formulated with almost all types of inhibitive pigments, metallic pigments (zinc, lead and aluminium), as well as lead salts, chromates and phosphates. Pigment volume concentrations of about 30 per cent are typical for all of these pigment types, although a proportion of extender material is normally included in the formulation. Where metallic zinc is used, however, P.V.C.s of about 65 per cent are required since the mode of protection of this pigment necessitates both intimate particle-to-particle and particle–substrate contact. An example of one of these 'zinc-rich' primers is given later.

An example of a brushing-grade chlorinated rubber–basic lead silico-chromate primer is shown in Table 4.17, and examples of a compatible white undercoat and a gloss finish are given in Table 4.18. The pigment volume concentrations of the primer, undercoat and finish in the three formulations are about 29, 35 and 16 per cent, respectively.

Thickeners. The use of hydrogenated castor oil as a thickener is a common alternative to the amine-treated clays but the incorporation techniques differ for the two additives. Hydrogenated castor oil requires high shear and high temperatures to effect gellation in a small quantity of solvent prior to its addition to the bulk paint, while the clay necessitates the use of a specific

Table 4.17 Lead Silico-chromate Primer

	Weight (%)
Lead silico-chromate	16·3
Titanium dioxide, rutile	5·0
Barytes	9·8
Chlorinated rubber, medium-viscosity grade	17·0
Chlorinated paraffin	8·5
Hydrogenated castor oil	0·6
White spirit	10·7
Trimethylbenzene	32·1

solvent addition technique (see page 33). However, similar rheological characteristics can be obtained with both materials.

Solvents. The solvent blend in the formulation given in Table 4.17 again contains white spirit, which is not a solvent for the chlorinated rubber polymer. However, provided it is not used in excess, the stability of the paint is not adversely affected, while its presence gives an improvement in brush application due to the slowing-down in the drying rate. Alternatively, a mixture of trimethylbenzene, white spirit and 2-ethoxyacetate in the ratio of 5:4:1 can be used; this is useful where low levels of aromatic solvents are necessary.

High-build chlorinated rubber paints

High-build chlorinated rubber paints can be formulated for either brush or airless spray application and these systems can produce dry film thicknesses of the order of 100–200 μm, typically over the lower-build primers of the type shown in Table 4.17. Alternatively, a low-build finish (Table 4.18) can be used cosmetically over a high-build base coat.

Table 4.18 Chlorinated Rubber Topcoats, White

	Undercoat weight %	**Gloss finish** weight %
Titanium dioxide, rutile	22·6	16·5
Mica	8·0	–
Talc	4·0	–
Chlorinated rubber, medium-viscosity grade	15·3	19·5
Chlorinated paraffin	10·2	13·0
Hydrogenated castor oil	0·3	–
Trimethylbenzene	29·7	38·3
White spirit	9·9	12·7

These high-build coatings are based on low-viscosity chlorinated rubber, generally in combination with a mixture of solid and liquid chlorinated paraffin plasticisers, as opposed to the more common practice of using only liquid plasticisers. Ratios of binder to plasticiser are generally 1:1 and the ratio of solid to liquid plasticiser is usually 2:1. The lower-viscosity grades of rubber are used, since they tend to minimise micro-void (bubble) formation during airless spraying while maximising the resin solids content for any required paint viscosity.

Pigmentation. For marine applications involving total immersion of the painted structure, high-build coatings with a pigment volume concentration somewhere in the range of 25–35 per cent provide best results; for other applications, levels down to about 18 per cent can be adopted. Inorganic pigments are generally favoured in these high-build coatings, usually in combination with barytes extender, which can comprise up to 50 per cent by weight of the total pigmentation. Permanent organic pigments can be used in conjunction with higher loadings of barytes when a cosmetic effect is required. However, in such cases cost considerations normally favour the use of organic pigments in low-build finishes subsequently applied over the high-build variant which has been pigmented with a suitable inorganic base.

Solvents. In brush-applied formulations, xylene can be used as the sole solvent, although some higher-boiling solvents such as trimethylbenzene or 2-ethoxyacetate can be incorporated to reduce the drying rate. The inclusion of a slower-evaporating solvent is essential in airless spray formulations since this tends to reduce micro-void formation in the film. Ratios of 4:1 for the xylene and high-boiling solvent mixture are suitable.

Table 4.19 High-build Finish for Steelwork, White

	Weight (%)
Titanium dioxide, rutile	14·3
Barytes	14·3
Chlorinated rubber, low-viscosity grade	14·5
Chlorinated paraffin, solid type	9·6
Chlorinated paraffin, liquid type	5·0
Clay gellant	2·8
Industrial alcohol	1·0
Xylene	30·8
2-Ethoxyacetate	7·7

The formulation of a high-build white chlorinated rubber paint for airless spray application is shown in Table 4.19. This system has a pigment volume concentration of about 30 per cent.

Two-pack systems
High-durability coating systems with very high chemical resistance and toughness can be prepared using two-pack epoxy or polyurethane resin systems.

Two-pack epoxy systems
The two-pack epoxies are available as both solvent-free and solvent-containing systems, of which the latter are preferred for use in the industrial and marine fields. In the two-pack epoxy systems, one component comprises the epoxy resin and the pigments while the other contains the curing agent, which is typically an amine, a polyamine, an amide or an adduct. When these are mixed together prior to application, the curing agent initiates the chemical reaction which ultimately results in the formation of the film. Film formation is generally considered to be effected by opening of the terminal glycidyl ether groups of the epoxy resin, followed by condensation polymerisation to yield a cross-linked resin.

Both semi-solid epoxy resins and low-molecular-weight solid products can be used in the solvent-borne coatings. The solid resins impart greater flexibility to the film while the semi-solid resins are more suited for high-solids formulations because of their better solubility characteristics.

The quantity of amine-containing curing agent required for use with an epoxy resin can be calculated from the relation:

$$A = E \times H$$

where A = amount of amine (g/kg epoxy resin)
E = epoxide equivalent per kg epoxy resin
H = hydrogen-active equivalent of the curing agent, i.e. the molecular weight divided by the number of amine hydrogens present in the molecule.

The value obtained by the above relation is the stoichiometric quantity of amine required to effect cure and it is normal practice to add 10 15 per cent excess of curing agent to ensure that the reaction proceeds to completion.

Solvents. Combinations of true solvents, to give the required evaporation rate and flow characteristics, and diluents, to reduce the raw materal costs, are used in these systems. In general, a blend of 40–50 per cent true solvent and not more than 50–60 per cent diluent is used. Glycol ethers and ketones can be used as true solvents and aromatic hydrocarbons as diluents. Esters, although good solvents for epoxy resins, tend to react with amine curing agents and thus can only be used in the epoxy component of two-pack systems.

Pigmentation. Epoxy resin paints can be pigmented with a wide range of pigment types: a selection criterion is that the pigments not only have to withstand the exposure conditions but must also be resistant to amine attack. The presence of relatively strong solvents also necessitates care in the selection of organic pigments. Suitable pigments include titanium dioxide, iron oxides, most inhibitive pigments, phthalocyanine colours and carbon black. Extenders such as barytes, talc and mica are used to impart specific properties to the film and rheological control can be obtained through the use of amine-treated clays.

Primers. An important primer type frequently based on an epoxy resin binder is the zinc-rich primer, Table 4.20. This type of primer is widely used on structural steelwork for bridges, etc., and as a primer for steel plate in shipyards. However, zinc-rich primers are only really effective if applied to blast-cleaned steel, preferably immediately after the cleaning operation has been performed. The zinc metal content of these primers ideally needs to be such that 92–95 per cent of the dried film is composed of metal, to establish the necessary particle-to-particle and particle-to-substrate contact essential for the sacrificial protection this type of primer provides.

Zinc-rich primers are manufactured using relatively simple stirring techniques, and a whole variety of alternative binders can be used. These alternatives include chlorinated rubber, isomerised rubber, polystyrene and inorganic silicates. With inorganic silicates the primers have to be in two-pack form as with the epoxies described above; the other binders all produce one-pack zinc-rich primer formulations.

Two-pack polyurethane systems

The polyurethane paints for structural steelwork are, in a similar manner to the epoxy resins, capable of providing highly durable coating systems for severe exposure conditions. Also, like the epoxy systems, these solvent-borne formulations are normally prepared as relatively low-film-build components with the paint system typically comprising a primer and two coats of topcoat.

Binders. The two-pack polyurethane systems are typically based on isocyanate adducts derived from the reaction between a di-isocyanate and trimethylolpropane or a polycyanurate, these adducts being reacted with a polyol to yield the final cross-linked resin. The polyol component is a resin containing free hydroxy groups, for example a polyester or a polyether.

In common with all high durability systems, polyurethane-based coatings need to be applied to steel which has been thoroughly prepared in order to obtain maximum performance. For most applications on steelwork, highly branched resins and polyfunctional isocyanates are used in order to produce a highly polymerised, impermeable film. Where extreme climatic changes might be expected, part of the resin component may be replaced with one having a degree of branching sufficiently low to impart flexibility to the resultant film, which can then accommodate any thermally-induced movement.

In hydroxy-cured polyurethanes the rate of cure is partly governed by the

Table 4.20 Two-pack Zinc-rich Epoxy Primer

	Weight (%)
Epoxy Component	
Zinc dust	83·0
Surface-coated calcium carbonate	1·0
Solid epoxy resin, diphenylolpropane based. Epoxy equivalent about 500	3·5
2-Ethoxy acetate	3·5
Methyl isobutyl ketone	2·9
Xylene	2·9
Curing Agent	
Polyamide resin, 100% non-volatile liquid. Amine value about 220	2·0
Methyl isobutyl ketone	0·6
Xylene	0·6

reactivity of the polyol: the more highly branched materials give faster drying times than the substantially linear polyols. The isocyanate components also vary in reactivity, with similar modifying effects on the curing time of the system and, for example, aromatic isocyanates have faster reaction rates than the aliphatic isocyanates.

For topcoats, combinations of any type of polyisocyanate and polyol can be used; for primers, combinations of a polyether (which gives maximum chemical resistance) and diphenylmethane di-isocyanate are favoured. Irrespective of the particular combination chosen, however, the yellowing tendencies of all isocyanates are pronounced, particularly the diphenylmethane di-isocyanate types.

The amount of isocyanate required to react with a given quantity of the polyol component is determined from a knowledge of the respective isocyanate value and hydroxyl value of the materials used. It is most convenient, however, to obtain this information from the resin manufacturer's data sheets, as is the case for two-pack epoxy systems. Nevertheless, improved performance can often be obtained by using up to 25 per cent more isocyanate than calculated for a 100 per cent (stoichiometric) cure.

Film-forming additives. Other film-forming additives can be added to polyurethane systems to modify the performance characteristics. In metal primer formulations the two most commonly used additives are castor oil, which upgrades water resistance (although this is achieved at the expense of some oil and solvent resistance), and epoxy resins, which can improve adhesion and upgrade chemical resistance. Solid epoxy resins of low molecular weight are used where flexibility is important, whereas the more highly condensed products are used where maximum chemical resistance is required. The epoxy addition is especially useful in metal primers since the preferred isocyanate type for this application, diphenylmethane diisocyanate, produces highly acidic byproducts. The epoxy resin, besides usefully modifying the film properties, functions as an acid-acceptor and therefore enables this isocyanate type to be used on both ferrous and aluminium substrates.

Solvents. Solvent systems for the two-pack hydroxy-cured polyurethanes generally consist of mixtures of esters and ketones, such as *n*-butyl acetate, ethyl glycol acetate (or 2-ethoxy acetate), cyclohexanone and methyl ethyl ketone. Aromatic hydrocarbons can be used as diluents, although certain isocyanates are totally incompatible with them. Diluents, even if compatible, have the effect of reducing the pot life of the two-pack systems when mixed.

The solvent mixture adopted has to be matched to the intended application technique. For example, brushing systems would be based mainly on high-boiling solvents such as ethyl glycol acetate and cyclohexanone, while a very high-boiling solvent may be included to improve the flow and wet-edge time of the film. For spray application, medium-to-low boiling solvents such as butyl or ethyl acetates and methyl ethyl ketone are used, as well as some higher-boiling material to improve flow. Solvents for the isocyanate component must be free of water in order to minimise the risk of gellation during storage.

Pigments. Pigments for use with these systems have to be selected with some care since certain types tend to react chemically with the isocyanate component and reduce the pot life of the mixed components. Examples of such pigments are certain zinc and lead chromates, zinc oxide, red lead, chromium oxide, Prussian blue, asbestine and some silicas, and all basic pigments. Despite their chemical reactivity, many of these pigments and extenders can still be used provided the pot life obtained is judged to be acceptable for the particular application. All pigments, however, are preferably dispersed into the polyol component as the water which is always absorbed on pigment surfaces tends to cause progressive gellation.

Organic pigments may adversely affect flow (which usually tends to be poor for the two-pack polyurethanes) and gloss; consequently, where required, these pigments are used at low addition levels.

Combinations of zinc chromate and red oxide of iron are used in primers for steel. For aluminium, zinc chromate comprises the whole of the pigment. In both instances, however, extenders are also used.

The large number of components that can be used in the preparation of polyurethane-based primers make it difficult to present a typical formulation. However, in Tables 4.21 and 4.22 outline formulations of a

Table 4.21 Two-pack Polyurethane Primer for Steelwork

	Weight (%)
Pigmented base	
Red oxide of iron	5·0
Zinc chromate	5·0
Barytes	16·5
Talc	8·2
Asbestine	8·2
Polyether resin, 90% in butyl acetate, hydroxy value about 300 mg KOH/g	3·6
Castor oil	6·2
Solid epoxy resin, diphenylol-propane based. Epoxy equivalent about 500	1·2
Cyclohexanone	25·1
Methyl isobutyl ketone	6·3
Curing agent	
Diphenyl methane di-isocyanate, technical. 60% in xylene Isocyanate value 18	14·7

metal primer for steel and a white gloss topcoat are given. The pigment volume concentrations are about 40 per cent and 16 per cent respectively, and the primer formulation in Table 4.21 demonstrates the use of castor oil and epoxy resin modifiers. The silicone fluid included in these formulations functions as a flow improver, and typically the addition level is at about 0·02 per cent silicone on the formulation.

Table 4.22 Two-pack Polyurethane Gloss Finish, White

	Weight (%)
Pigmented base	
Titanium dioxide, rutile	14·6
Polyester resin, 80% in butyl acetate, hydroxy value about 175 mg KOH/g	18·1
1% solution silicone flow additive in ketone solvent	0·2
Cyclohexanone	4·0
Methyl ethyl ketone	30·5
Toluene	12·6
Curing agent	
Toluene di-isocyanate adduct, 40% in butyl acetate, isocyanate value about 6	20·0

4.5 Bibliography

Plaster, H.J., *Blast Cleaning and Allied Processes*, vols I and II. Portcullis Press: London, 1972.

Newton, D.S. (Ed.), *Paint Technology Manual*, Part 4, *The Application of Surface Coatings*. Chapman and Hall: London, 1965.

Tatton, W.H. and Drew, E.W., *Industrial Paint Application*. Newnes: London, 1964.

Murphy, J.A. (Ed.), *Surface Preparation and Finishes for Metals*. McGraw-Hill: London, 1971.

Von Fraunhofer, J.A. and Boxall, J., *Protective Paint Coatings for Metals*. Portcullis Press: Redhill, 1976.

Flick, E.W., *Solvent-based Paint Formulations*. Noyes Data Corporation: New Jersey, 1977.

CHAPTER 5

Formulation of Coatings for Timber

5.1 Introduction

The use of timber as a structural material, and for fittings and components, is widespread in the building industry. Softwoods are the most widely used timbers in the construction industry; hardwoods and board products are used in the furniture industry as well as for specific building applications.

The terms 'softwood' and 'hardwood', although strictly non-technical, are widely recognised and used to differentiate between the two main botanical classes of timbers. Softwoods are obtained from cone-bearing species which have evergreen needles, whereas hardwoods are obtained from broad-leaved trees. The structure of the hardwoods is more complex than that of softwoods and their properties show a much wider range of variation.

The method of growth, however, is identical for both classes of timber, nourishment being by the respiration of oxygen and carbon dioxide from the air and ground water. The oxygen, carbon dioxide and water are converted into starch and sugars, which are in turn synthesised to make wood substance.

Structure of timber

A cross-section through a tree reveals three outer layers, namely bark, which provides protection, phloem, which transports the food, and the cambrium, which is the growing zone of the tree where all new cells are formed. The next (inner) zone is the xylem, which is the wood component, and there is a central small area of pith.

The xylem consists mainly of fibres (or tracheids), which are the main strength element of the 'wood', and cells forming the rays. In hardwoods, elements known as vessels or pores are a major anatomical feature.

Cross-sections of tree trunks of both hardwoods and softwoods grown in temperate climates show annual growth rings. These are made up of larger cells formed in the spring when growth is at a maximum, earlywood, and latewood which is formed in the summer and whose cells are denser and harder. The outermost growth rings are collectively known as 'sapwood' and act as a food storage and water transport zone, whereas the inner, older growth rings form an inactive zone solely providing strength and known as 'heartwood'.

The xylem (wood) consists mainly of fibres or tracheids which are the main strength-providing elements; these are long hollow cells which run vertically in the tree. In the cell wall of these elements are bordered pits which facilitate the movement of solutions from one cell to the next.

Rays are another major feature of the system and these, formed from strips of cells, run more or less horizontally from the bark to the pith, at a right angle to the direction of the fibres. These cells are connected by pits in the cell walls, and, in the sapwood, this facilitates transport of food. In addition to these pits and the bordered pits, both of which allow transport in the cells and fibres respectively, half-bordered pits also exist in the common walls between cells and fibres. The prime importance of these half-bordered pits is that, in both living and dead wood, fluids and finely dispersed solids may pass through them.

In hardwoods, while both the structural fibres and the cells are present, a characteristic feature is the presence of vessels which run longitudinally, i.e. parallel to the trunk, and which conduct water. The distribution of these vessels, which may be quite large depending on the species, varies and results in the broad subdivision of hardwoods into the categories known as 'ring-porous' and 'diffuse-porous'. When viewed in cross-section, the vessels in ring-porous hardwoods are grouped into circular zones, while those in diffuse-porous hardwoods are randomly distributed.

Chemical properties

Chemically, wood is composed of about 50 per cent cellulose and approximately 25 per cent each of lignin and hemi-cellulose. As a result of its crystalline, filamentous nature, cellulose provides the high tensile strength of timber. The lignin and hemi-cellulose serve as a matrix cementing together the cellulose filaments and imparting strength and rigidity to the cell walls. Certain minor constituents, such as various gums, resins and dyestuffs, may also be found within cells and these minor constituents can have a disproportionate effect on coating performance. Many of the properties of timber are dependent upon the chemical composition, perhaps most importantly its natural durability on weathering.

Several major industrial products are obtained from wood, notably paper and pulp, rayon, cellulose polymers for coatings, wood alcohol, turpentine and tanning materials.

Moisture content

Timber for use in construction and industry has to be seasoned. This drying process involves the loss of moisture from the cell cavities and the cell walls, and thus shrinkage. In practice, the moisture content is reduced to a point where it is in equilibrium with local conditions. For exterior use and structural purposes, e.g. roofing timbers or cladding, a moisture content in the range of 14–22 per cent is required—this is the so-called 'air-dry' timber. Moisture contents of 12 per cent or less are required for timbers for interior use so that the timber is in equilibrium with the room.

Because of the hygroscopic nature of wood, seasoned timber absorbs moisture and swells when exposed to dampness but it will dry and shrink again when drier conditions prevail. This effect, known as movement, which has to be accepted as a natural characteristic of timber, has major implications for the paint system performance, especially on exterior exposure.

Pretreatment

The perishable nature of sapwood necessitates that many timbers for exterior use are pretreated against fungal decay. This involves immersing the timber in a preservative solution which, typically, is solvent-borne tri-*n*-butyl tin oxide for dimensioned components or water-borne copper–chrome–arsenic salts for non-dimensionally critical constructional timbers. Fuller details of timber pretreatments can be obtained from the specialised monographs listed in the bibliography at the end of this chapter.

5.2 Paint systems for timber

Coatings used on timbers can conveniently be subdivided on the basis of whether the service exposure conditions are exterior or interior. For exterior use, alkyd-based paint systems and exterior wood stains, which can also be based on alkyd resins, are widely used. The comparatively recently introduced alkyd-based exterior wood stains have largely superseded the clear varnishes previously used. For interior applications, such as doors, skirtings and windows, the use of alkyd-based paints identical to the systems used for exterior timbers predominates. Much industrially finished timber, such as furniture, cabinets and toys, is coated with systems based on polyesters or alkyd–amino combinations, or with nitrocellulose coatings.

Paint penetration

The successful formulation of durable coatings for exterior timber is one of the most difficult tasks facing the paint technologist. Paint performance on timber is largely dependent upon the penetration (and subsequent adhesion) of the coating system into what is both a chemically and a physically complex material and, accordingly, a discussion of penetration is of value here.

In softwoods, the penetration of pigmented paints and clear varnishes is determined by the rays and the wood fibre. If ray cells are open on the surface, penetration will occur further into the cells by flow through the pits in the ray walls. This process can continue into the cavities of adjoining tracheids by means of the half-bordered pits in the common cell walls between the ray cells and the tracheids. The penetration of paints can be very deep in ray cells, although normally it is limited in an 'outward' direction to a few of the tracheids adjoining the rays.

Penetration can also occur into gross checks in weathered wood and also into the splits caused by the degradation by ultra-violet light of the 'cement', known as the middle lamellar, that binds all cells in wood together.

There are, however, some differences in behaviour between springwood and summerwood. In particular, the somewhat larger cell cavities in springwood allow greater penetration than that which occurs in summerwood and this can result in uneven film thickness and gloss levels. Furthermore, with pigmented systems, both the pigment and binder have relatively easy access into the cells of springwood, whereas with summerwood a selective filtering of the pigment particles occurs, with only a part of the binder penetrating the cells. These effects over growth bands have a tremendous influence on the subsequent performance of paint systems during service, and it is often found that coating failure occurs preferentially over summerwood growth bands.

In hardwoods, the ray cells are not important in paint penetration processes although the reason for this is not known. The major route of entry is into the vessels, where both pigment and binder can flow, and the tracheids, into which only the binder flows. The vessels in many species of hardwood are very wide and this allows ready penetration of coatings. However, this can result in the formation of a thin, stressed film over the sharp edges of the vessels at the surface, and the rather poor coating performance on timbers such as oak, iroko and afzelia can often be attributed to failure initiated at vessel openings.

In other hardwoods, however, the structure is such that penetration into vessels is not excessive. In some instances this is due to physical blocking by naturally occurring resinous components. Although this can result in an improved appearance and performance, too high a loading of resins and extractives can cause problems with adhesion and paint-drying, respectively.

To obviate the possible adverse effects of the structural features of timbers on paint performance, various techniques and processes are employed. With hardwoods, the use of fillers is common; these physically block the open ends of vessels and artificially unify paint flow and penetration. Timbers with known high extractive contents, which can cause lengthening of the drying time of oxidative-cured systems, are sometimes given a prior coat of moisture-cured polyurethane sealer.

With softwoods, sanding can close up the wider springwood cells and, consequently, the use of fillers is often not required. High viscosity primers can also be used, and these would be expected to reduce the flow into cavities. The use of emulsion-based systems also results in reduced penetration since, in this instance, the particles of the dispersed resin are too large to penetrate the cells.

5.3 Coatings for exterior timbers

5.3.1 TYPES OF PAINT SYSTEM

The alkyd-based paint formulations used on exterior timbers are generally applied as three-coat systems comprising primer, undercoat and, normally, a full gloss finish. Application is almost always by brush, although frequently a factory-applied primer coat is used on joinery components for new buildings, and this is generally applied by immersion or deluging techniques, or sometimes by spray application.

The components of these paint systems have traditionally been based on binders which cure by air-drying (i.e. oxidative processes). Recently, emulsion-based systems which dry by water-loss and coalescence have begun to play an increasingly important role.

The composition and formulation of the alkyd-based undercoats and topcoats used on exterior timbers are identical to those alkyd systems used on steelwork (page 108).

5.3.2 PRIMERS

Various types of primer are available for exterior timbers. They may be broadly categorised into lead-containing, lead-free, emulsion-based and aluminium-based.

Lead-based primers

The use of lead-based primer on exterior timber is now rare because of the toxicity risk associated with lead-containing paints, although in terms of performance these primer types have never been bettered. The formulation of a lead-based wood primer is shown in Table 5.1.

The pigment combination, a blend of white lead and the non-setting grade of red lead in the ratio 9:1, tends to produce a degree of film elasticity (due to soap formation with the oil binder) so that dimensional movements in the timber can be accommodated. The resultant colour from this particular combination gives this type of primer its common name of 'pink primer'.

The use of a blend of linseed oils, the low-viscosity alkali-refined linseed oil and the high-viscosity linseed stand oil, results in films of very good adhesion and integrity. This is largely a result of the tendency of the

Table 5.1 Lead Primer for Timber

	Weight (%)
White lead	63·9
Red lead	7·1
Surface-coated calcium carbonate	3·0
Alkali-refined linseed oil	15·0
Linseed stand oil, 100 poise	5·0
White spirit	5·4
24% Lead naphthenate	0·4
6% Cobalt naphthenate	0·2

low-viscosity components to penetrate into the timber, while the higher-viscosity oil is retained substantially on the surface.

The formulation shown in Table 5.1 conforms to British Standard BS 2521, *Pink Primer for Woodwork*, which stipulates that total linseed oil content should be between the limits of 17 and 20 per cent, with a maximum content of 6 per cent white spirit and 4 per cent suspending agent (asbestine, kieselguhr or a surface-coated calcium carbonate). The pigment volume concentration of the example shown is about 35 per cent. Despite relatively high levels of drier, the drying time of these lead-based primers tends to be fairly long and this, together with health and safety considerations, has restricted their present use.

Lead-free primers

Lead-free or, more correctly, low-lead primers are formulated on combinations of opacifying pigments and extenders, typically in conjunction with an oleoresinous varnish. Pigment volume concentrations tend to be in the region of 40 per cent, and white or pastel shades and, especially, pink (in allusion to the pink lead primers, and obtained by the addition of a small amount of red oxide of iron to a white base) are the standard colours. Various extenders, usually whiting and barytes, are used in admixture with the colouring pigment, while low levels of talc or mica are often added to improve film integrity.

The binder types most frequently used are the oleoresinous varnishes, although alkyds offer advantages with respect to film flexibility and hence durability on wood. Low-lead primers are widely used in the factory priming of many timber components, especially window joinery. In this application, speed of drying is of primary importance and faster drying can be accomplished by using short-to-medium oil-length materials which tend to cure rather more rapidly than longer-oil-length binders. For on-site use, where the drying time is somewhat less important, long-oil-length binders can be used and this produces formulations of better durability.

Table 5.2 Acrylic Wood Primer

	Weight (%)
Titanium dioxide, rutile	17·5
Barytes	10·9
Whiting	7·2
Mica	3·9
Acrylic emulsion, 48% solids	43·5
Cellulose thickener	0·2
Water	15·1
10% solution sodium hexametaphosphate pigment dispersant	0·4
Bactericide	0·2
Defoamer	0·1
High-boiling ether coalescing solvent	1·0

Emulsion-based primers

An alternative type of low-lead primer is based on emulsion resin technology as described on page 148. Emulsion wood primers are almost invariably based on pure acrylic resins, which form films that are characterised by high flexibility and thus are ideally suited for use on a dimensionally unstable substrate such as timber. Furthermore, acrylic resins retain their flexibility on weathering as a result both of their high resistance to oxidation and of their inherent high stability towards ultra-violet light. A useful general characteristic of emulsion primers is their speed of drying, generally of the order of 1–2 hours under ambient conditions. This has advantages in both site and factory applications.

Pigment volume concentrations for emulsion wood primers are typically in the region of 30–40 per cent and, where the primer–undercoat type of formulation is required (two coats of the same material), higher P.V.C.s of about 45 per cent can be adopted. The pigmentation of acrylic emulsion primers presents few problems. The colour required is usually white or a pastel shade and combinations of titanium dioxide, whiting and barytes, with small amounts of mica or talc to strengthen the film and improve water resistance, are typically used. The formulation of a white acrylic primer for timber of about 37 per cent pigment volume concentration is shown in Table 5.2. This would be suitable for brush application. The functions of the various additives in emulsion paint formulations are summarised on page 148.

Aluminium-based primers

The last major type of wood primer is the aluminium primer, which tends to have more specific uses than those already described. Unlike the other low-lead primers, aluminium primers are only applied by brush under site conditions. They are favoured for application to hardwoods and to soft-

Table 5.3 Aluminium Primer for Timber

	Weight (%)
Non-leafing aluminium paste, 70% powder in white spirit	24·3
Medium–long-oil-length tung oil phenolic varnish, 70% in white spirit	36·5
White spirit	38·5
24% Lead naphthenate	0·5
6% Cobalt naphthenate	0·2

woods with many knots or resinous areas, and also for timbers prone to exudation of the natural wood resins, where they are thought to be beneficial in holding back the resins which might otherwise cause disruption of the applied coating system.

As their name suggests, the major component of the pigmentation in these primers is aluminium. The metal is used in the form of a fine flake, and the so-called non-leafing grades are preferred since the more random platelet distribution obtained at the film surface tends to improve the adhesion of the subsequently applied undercoat. Despite this, paint failure with systems based on aluminium primers may occur, frequently as a result of intercoat adhesion failure between the primer and its topcoats.

Binders are normally of the oleoresinous varnish type, and tung oil–phenolic varnishes as well as combinations of oils and the neutral coumarone or hydrocarbon resins are most frequently used. A major requirement of the binder, irrespective of its composition, is that its level of free acidity should be very low. Too high an acid value results in reactivity with the aluminium metal leading to the phenomenon of 'gassing', which can cause accidents and damage during storage of the paint in sealed containers.

Generally, aluminium is used as the sole pigment in these primers, although extenders can be included to modify film properties or to lower raw material costs. Pigment volume concentrations are typically in the region of 20–25 per cent. The formulation of such a primer of 20 per cent P.V.C. is shown in Table 5.3.

5.3.3 FINISHES

Exterior wood stains

Exterior wood stains occupy an intermediate position between pigmented systems on the one hand and clear varnishes (see later) on the other. At a very simple level, exterior wood stains can be considered as being relatively dilute varnishes that contain small amounts of pigments together with sufficient fungicide to minimise biological growth on the film during service. The pigment content of exterior wood stains can be varied to produce different cosmetic effects, and systems that are semi-transparent or opaque are used. Furthermore, the binder content can be adjusted to produce systems of either low ($<$about 30 μm) or high (about 50–80 μm) film-build in contrast to the film-build of about 100–120 μm for a conventional alkyd paint system.

With exterior wood stains the coating system is built up by a coat-on-coat application of the same product, usually in two or three coats. In this respect they clearly differ from conventional types of paint system where components of dissimilar composition are usual.

The main advantage of exterior wood stains is their cosmetic appeal, that is, they tend to enhance the natural appearance of timber rather than obscure it as does a conventional paint. Even opaque wood stains show grain patterning since the film is comparatively thin. A second and most important advantage is that correctly formulated systems do not exhibit the gross cracking and flaking failure characteristic of most paint systems, but tend to fail by erosion. This facilitates preparative work prior to maintenance since no labour-intensive scraping, sanding or burning-off is required.

Exterior wood stains are typically based on long-oil-length drying-oil alkyds. Although, as with paints, a variety of binder types can be used, alkyd or oleoresinous varnishes are favoured; however, formulations may be based on acrylic emulsion, or on combinations of alkyd and acrylic emulsion.

Pigmentation. Pigment volume concentrations for the semi-transparent stains tend to be in the region of 2–5 per cent. Where opaque systems are required, P.V.C.s of up to about 25 per cent may be used, and extenders such as mica or talc are frequently used in conjunction with the colouring pigments. Only completely light-fast colours are used in exterior wood stains and, furthermore, pigments that strongly absorb ultra-violet light are favoured since they tend to maximise protection, both of the polymer component and of the timber substrate. Yellow, red and brown oxides of iron and carbon black are the most widely used colours—the whole range of natural wood shades can be achieved with combinations of these pigments. Micronised iron oxide pigments are generally more effective than coarser material in these formulations.

The formulations of a low-build and a high-build semi-transparent exterior wood stain compared in Table 5.4 have pigment volume concentrations of about 5 per cent and 2 per cent, respectively. These stains would be brush-applied in two coats to timbers, although three coats would be preferred in more severe exposure conditions or where longer maintenance periods were required.

Table 5.4 Exterior Wood Stains

	Low-build (weight %)	High-build (weight %)
Red oxide of iron	5·6	2·0
Long-oil-length drying-oil alkyd, 75% in white spirit	33·3	48·6
Fungicide	0·2	0·5
White spirit	60·9	48·5
24% Lead naphthenate	—	0·3
6% Cobalt naphthenate	—	0·1

Binder. Binder content varies depending on whether a low-build or a high-build stain is required. Higher build produces lower moisture permeability and hence reduced dimensional movement of the substrate.

Drier. The use of driers in the relatively low-build systems is not normally necessary and, indeed, heavy metal driers are detrimental to film durability in this type of system since they tend to initiate the cracking and gross flaking type of failure. Higher-build products which would stay tacky for inordinately long periods can include a drier, although at levels of only about 50 per cent of the recommended level for the binder when used in a conventional paint system.

Other additives. Certain interesting formulation differences are apparent in exterior wood stains compared with conventional alkyd paint systems for timber. Fungicides are generally added to wood stains in order to provide protection against the disfiguring moulds and blue-stain fungi which can colonise films; several types can be used, such as organo-mercurials, chlorinated phenols and isothiazolin derivatives. Addition levels are obviously dependent upon the efficacy of the fungicide used.

Another major difference is that the viscosity of these systems is typically lower than that of paints, being designed to optimise penetration into the timber. Performance and appearance can be improved, however, if some restriction is placed on penetration, and this can best be achieved by using low addition levels of amine-treated clay gellants. The technique of thickening formulations with such materials is to stir the clay into a relatively large proportion of non-polar solvent (for example, white spirit), and then to add specific polar solvents (typically an alcohol) to cause gellation to occur. This so-called pre-gel can then be dispersed into the bulk formulation to control its rheology.

Emulsion-based systems

Many coatings used on timbers are of higher durability than the pigmented alkyd systems and the alkyd-based exterior wood stains, which rarely give

Table 5.5 Alkyd–Acrylic Emulsion Wood Stain

	Weight (%)
Red oxide of iron	1·5
Acrylic emulsion, 50% solids	30·0
Alkyd emulsion (Table 5.6)	40·5
Polyacrylate thickener, 10% aqueous solution at pH 9·0	10·0
Water	17·0
High-boiling ether solvent	1·0

more than four years of maintenance-free life. All acrylic emulsion-based paint systems, highly pigmented alkyd paints applied in two or three coats, as well as acrylic emulsion-based stains or alkyd emulsion-based stains, are among the formulation types potentially capable of providing better performance.

Acrylic emulsion paint systems are comprised of acrylic-based primer–undercoats and a compatible acrylic gloss or, more frequently, a semi-gloss finish. A semi-gloss finish can be formulated along similar lines to the primer shown in Table 5.2, but with a pigment volume concentration of 18–20 per cent composed entirely of colouring and opacifying pigments.

The modified alkyd systems are formulated with pigment volume concentrations of the order of 20–30 per cent, using colouring pigments and extenders. Other formulation details, such as volume solids, solvents, binder type, driers, etc., are similar to the alkyd gloss paints (Table 4.12, page 109), but the system is composed of just two coats of the 'finish'. Higher durability is achieved by the omission of the highly pigmented alkyd undercoat (Table 4.13, page 109), which is invariably a source of weakness in coating systems on timber, by the lower film thickness which allows a breathing action, as well as by a reduction in the tendency towards embrittlement and flaking by means of the higher pigment volume concentration. These systems differ from the opaque wood stains in having somewhat greater film build and much higher viscosity, and in that there is no attempt made to enable the grain pattern of the timber to 'signature' through.

The alkyd emulsion-based systems are invariably formulated as high-build wood stains, with a low pigment volume concentration. A compromise of properties is obtained with these systems: the alkyd penetrates the timber, taking any fungicide with it, while the acrylic emulsion is retained on the surface with the pigment. Generally applied as two-coat treatments by brush, these systems are far more durable than either straight alkyd or acrylic emulsion formulations.

The formulation of a semi-transparent alkyd emulsion wood stain shown in Table 5.5 has a pigment volume concentration of about 1·5 per cent. The

1:1 ratio of acrylic to alkyd solids shown in this example can be considered optimum for durability, although ratios ranging from 1:2 to 2:1 can also be used successfully.

Manufacture. The method of manufacture of these systems is relatively complex and involves emulsifying the alkyd resin component into an aqueous base using ideally, although not necessarily, liquid alkyd resins of 100 per cent non-volatile contents. Typically, the pigment is dispersed with part of the water and the polyacrylate thickener solution using a high-speed stirrer. The alkyd emulsion and then the acrylic emulsion are run in under slow speed, followed by the balance of water and the coalescing solvent. The alkyd emulsion (Table 5.6) is prepared by premixing the alkyd resin, driers and surface-active agent with a high-speed stirrer. The premixed water and ammonia are then added slowly while stirring; once a smooth paste has been achieved, the final water addition is made. Stability of these alkyd–acrylic emulsion systems is difficult to obtain and considerable experimentation, especially with regard to the surface-active agent and formulation pH, is required to obtain good long-term results.

Polyacrylate thickeners, as used in Table 5.5, are commonly included in high-performance emulsion paint systems. It is normal to prepare a pre-gel solution by stirring the special polyacrylate material (in its 'as-supplied' emulsion form) into a proportion of the water. On addition of carefully controlled amounts of ammonia, the emulsion changes to a solution and thickens. The required amount of paste is then added to the paint formulation, generally at the stage when the pigments are dispersed.

These alkali-soluble acrylic thickeners are less water-sensitive than the more widely used cellulosic types; furthermore, they impart superior flow properties and thus they can also be used in gloss emulsions.

5.4 Coatings for interior timbers

The coating of interior timbers can be accomplished using any of the paints and stain systems discussed in the previous section and, indeed, most joinery, doors, etc., are coated with the conventional alkyd-based paint systems. For many interior applications, however, clear finishes are used and these can be broadly categorised into systems that the householder or hobbyist might use, and the rather more complicated coatings that would be applied under factory conditions to furniture, cabinets and other timber articles.

5.4.1 VARNISHES AND LACQUERS

The main clear finishes applied to internal timber substrates are the traditional varnishes which have been used for centuries for this purpose. In

Table 5.6 Emulsified Alkyd Resin

	Weight (%)
Long-oil-length drying oil alkyd, 75% in white spirit	49·3
6% Cobalt naphthenate	0·18
24% Lead naphthenate	0·42
Alkyl phenol ethylene oxide condensate, emulsifying aid	1·7
Water	27·3
Ammonia, s.g. 0·880	0·5
Water	20·6

its simplest form, a varnish can be considered to be a solution of either natural or synthetic resins in an appropriate solvent, without any colouring pigments being present. Varnishes are brush-applied as three- or four-coat treatments, coat on coat, without any special sealer or primer coat being used, although the first coat can sometimes be usefully thinned in order to assist penetration.

The many binder types that can be used in varnishes for timbers include tung oil-modified phenolics (the so-called marine varnishes), alkyd resins and polyurethane alkyds (uralkyds). The formulation of an interior varnish for woodwork based on alkyd resin is shown in Table 5.7. Medium-oil-length or even short-oil-length materials are used for specific purposes where rapid hardness is required to be developed, as for example with varnishes for seats and chairs.

The long-oil-length varnishes are still used on both interior and exterior timbers, although for the latter use they have the serious disadvantage of rapid failure induced by ultra-violet light. Latterly, however, techniques of modifying these varnishes to render them suited for exterior use have been developed. It is possible, for example, to improve resistance to ultra-violet light by incorporation of special organic stabilising additives such as

Table 5.7 Interior Varnish

	Weight (%)
Long-oil-length soya oil alkyd, 70% in white spirit	85·0
White spirit	12·3
6% Cobalt naphthenate	0·5
24% Lead naphthenate	1·2
Methyl ethyl ketoxime, anti-skin agent, 10% solution in white spirit	1·0

benzotriazole at about 5 per cent by weight on binder solids. In this manner, durabilities similar to or better than conventional pigmented alkyd paint systems can be achieved. Alternatively, inorganic extender materials with high ultra-violet absorbance can be incorporated into the formulation; provided that their refractive index matches that of the dried binder film, optical transparency within the film will be maintained. Materials such as pulverised fuel ash (a power station byproduct), mica and pumice can be used successfully in this manner with most alkyds and varnishes at pigment volume concentrations of about 10 per cent. These varnishes also have a modified failure mode compared with conventional systems, in that they tend to fail by progressive erosion which simplifies subsequent maintenance.

Notwithstanding this, clear varnishes are not widely used out of doors, having largely been superseded by the formulations described in the previous section. They can be used, however, quite successfully in interior situations where the problems of degradation under ultra-violet light are considerably less.

Catalytically-cured systems

Finishes applied under factory conditions need to be relatively quick-drying and normally this is achieved by the use of catalytically-cured systems (high-temperature stoving cannot be used on coated timber in view of stability considerations), although for some applications nitrocellulose-based systems are used which dry rapidly by solvent loss.

Foremost among the catalytically-cured systems are the acid-catalysed alkyd–amino and the peroxide-cured polyester types.

Alkyd–amino wood finishes

Alkyd–amino wood finishes are generally composed of combinations of short-oil-length non-drying oil alkyds and butylated urea-formaldehyde resins. Although butylated melamine-formaldehyde resins tend not to react with acids unless heated, specifically modified melamines are available for acid-cured systems and these can be used where greater resistance to boiling water may be required. One- and two-pack variants are used and can be applied by spray methods or curtain coaters.

Some basic formulation characteristics of alkyd–amino finishes have been discussed on page 95, although certain modifications are obviously necessary. With these types of wood finish, cure is effected by the use of an acid catalyst. In the case of two-pack types, mineral acids such as hydrochloric or sulphuric acid are used, when the pot life obtained can range from a matter of hours up to several weeks after mixing. The single-pack formulations contain organic acids such as acid butyl phosphate and, while these remain stable for periods of up to one year, the curing times on application are approximately 2–3 times longer than for the two-pack types.

Low addition levels of the acid catalysts are normally used, about 0·5 per

Table 5.8 Two-pack Alkyd–Amino Finish, White

	Weight (%)
Base	
Titanium dioxide, rutile	20·0
Butylated urea-formaldehyde resin, 65% in *n*-butanol	17·0
Short-oil-length soya alkyd, 55% in xylene	25·0
Acetone	15·0
Xylene	10·0
n-Butanol	11·5
Acid catalyst	
Sulphuric acid, 10% solution in *n*-butanol	2·5

cent for the two-pack and about 3 per cent for the one-pack types; obviously some control on the pot life of the formulation can be exercised by varying this amount—that is, lower levels result in longer pot life. Furthermore, increasing the amount of alkyd resin and/or its oil length has a similar effect, although the film hardness will be affected by such changes.

The solvent blend of alkyd–amino wood finishes can be modified from the xylene-*n*-butanol combinations which are normal for stoving-cured metal finishes. In particular, the use of the faster-evaporating alcohols as well as ketone solvents results in more rapid drying.

Where pigmented systems are required, the basic selection criterion is resistance to the acidic conditions prevailing in the formulations: clearly carbonates must never be used, while the chromates are variable in their resistance. Pigment volume concentrations of about 15–20 per cent are typical, the formulation in Table 5.8 having a P.V.C. of about 18 per cent.

Table 5.9 One-pack and Two-pack Alkyd–Amino Wood Lacquers

	One-pack (weight %)	Two-pack (weight %)
Short-oil-length dehydrated castor oil alkyd, 50% in xylene	36·8	39·0
Butylated urea-formaldehyde resin, 65% in *n*-butanol	42·4	44·5
Xylene	8·6	10·0
Methyl ethyl ketone	7·0	2·0
n-Butanol	2·4	2·0
Acid butyl phosphate	2·8	—
Hydrochloric acid, 20% in *n*-butanol	—	2·5

Formulations of one- and two-pack clear alkyd–amino lacquers are shown in Table 5.9. The general resistance properties of the one-pack types of alkyd–amino finish are somewhat inferior to the two-pack type because of the lower degree of cross-linkage that occurs during curing. Small additions of cellulose nitrate or vinyl copolymers can improve hardness levels of the film.

Polyester wood finishes

Polyester wood finishes are based on unsaturated polyester (alkyd) resins which are typically prepared from propylene glycol and maleic anhydride, no modifying oil being present. The solvent used, styrene, is fundamental to these systems since, on addition of the peroxide catalyst, the styrene cross-links with the polyester to form films of essentially 100 per cent non-volatile content. No other modifying solvent is required, however, since during application, i.e. before gellation, the styrene fulfils this function. High-build one-coat finishes can therefore be deposited from these formulations (Table 5.10).

Table 5.10 Polyester Lacquer

	Weight (%)
Base	
Polyester resin, 60% in styrene	80·0
Wax, 1% in styrene	5·0
Colloidal silica	1·0
6% Cobalt naphthenate	0·5
Styrene	11·5
Catalyst	
Methyl ethyl ketone peroxide	2·0

The curing of polyester systems is inhibited by oxygen, and the function of the wax additive is to float to the surface of the film prior to gellation to form a barrier to the atmosphere. After curing, this wax layer which matts the film is removed by sanding and full gloss is developed by a polishing process. This obviously limits the use of this type of finish to flat panel products, although chemically-modified polyester resins are available which can produce gloss finishes directly on application. The wax is also beneficial in reducing loss of styrene which would result in imperfectly cured films.

The colloidal silica addition is a thixotropic agent specifically required to eliminate the sagging that occurs as a result of the high film-builds obtained, although this is only required for systems spray-applied onto vertical surfaces. When colouring pigments are required, addition levels of 10–15 per cent pigment volume concentration are normally adequate because of the high film-build deposited; certain organic pigments, however, can interfere with the curing mechanism.

Manufacture and application. The cobalt naphthenate and peroxide additives control the curing of the polyester and styrene components. It is essential that the cobalt naphthenate and the peroxide are always isolated from one another since mixing them together can cause an explosion.

The cobalt naphthenate is always added to the resin component of the system; levels of 0·01–0·1 per cent metal on polyester solids are normal. The peroxide catalyst is added prior to application (at levels of 2–4 per cent on total weight); a pot life of the order of 15 minutes to one hour is typically obtained. This short pot life necessitates the use of metered twin-feed application techniques for certain uses, or alternatively the so-called 'active ground' system can be used. This involves the prior application to the timber of the peroxide carried in a nitrocellulose formulation. After drying, a topcoat composed of the polyester, styrene, wax and cobalt initiator can be applied and curing is induced by stoving at a moderate temperature. Much higher levels of peroxide (about 10 per cent) have to be used in the ground-coat method, however, since not all of it is available for reaction with the subsequently applied polyester. Nevertheless, this technique entirely obviates storage and pot-life problems.

Certain difficulties can be encountered when these peroxide-cured polyesters are applied to timbers. In particular, naturally occurring phenolic components from the timber can decompose peroxides and so cause poor film curing; also, with the thixotropic polyesters, penetration may be poor. To minimise these problems, polyesters are frequently applied over two-pack polyurethane sealers.

Nitrocellulose finishes

The use of nitrocellulose lacquers on many larger factory-finished articles has largely been superseded by the use of the higher-solids, lower-flammability alkyd–amino and polyester finishes described earlier. However, nitrocellulose paints and lacquers still find an important outlet in the finishing of small articles such as toys and small domestic items, and the general principles of their formulation can usefully be discussed here. It must be emphasised, though, that nitrocellulose formulations are among the most varied and complex likely to be met by the paint technologist, and the following is only a guide to the basic characteristics of certain of the constituents.

The main components of a nitrocellulose paint are the cellulose binder itself, which imparts strength and hardness, and a hard resin, typically ester gum or a natural resin, that increases the gloss level, solids and adhesion but which reduces tensile strength and exterior durability. Plasticisers are also added to increase gloss, durability and elasticity, although generally at the expense of hardness and tensile strength. Non-drying short-oil-length alkyds are used widely as plasticisers for nitrocellulose, although the chlorinated paraffins are also employed for this and have the advantage of reducing the flammability of the film. Various phthalates are also commonly used as plasticisers.

Pigmentation. The pigments used should impart high opacity and have a high tint strength, while low oil absorption is also desirable in view of the characteristic low solids content of nitrocellulose paint. Because of the strong solvents required in nitrocellulose formulations, the 'bleeding' characteristics of organic pigments must be low to prevent staining-through as successive coats are applied. This criterion largely eliminates the Hansa yellows, Helio reds and maroons. If required, extenders can be used; as with pigments, low-oil-absorption types are preferred. Special extenders are used for specific purposes, for example, slate powder in metal primer–surfacers, and silica to control the gloss level of clear lacquers. One specific additive, zinc stearate, is used as a flatting agent in sanding sealers for wood.

Table 5.11 Cellulose Sealer for Timber

	Weight (%)
Cellulose nitrate, low-viscosity grade, damped 30% with alcohol	10·0
Ester gum	8·0
Dibutyl phthalate	4·0
Zinc stearate (flatting agent)	2·0
Butyl acetate	10·0
Ethyl acetate	10·0
Isopropyl alcohol	10·0
Toluol	46·0

The dispersion of pigments in nitrocellulose paints is best accomplished in ball mills since this reduces solvent loss to a minimum. Irrespective of the dispersion technique used, overheating must be avoided since this can decompose the nitrocellulose polymer.

Solvents. The solvents for nitrocellulose are capable of wide variation to suit application techniques and to control drying, flow, flash-point, bleeding characteristics of pigments and raw material costs. The solvent compositions can be subdivided into three classes: active solvents, latent solvents and diluents.

Active solvents, such as esters and ketones, are true solvents for nitrocellulose. Latent solvents are not true solvents but improve the solvating power of the active solvents; alcohols are typical examples. The diluents are not solvents for nitrocellulose itself although they will dissolve the secondary components, the hard resin and plasticiser, and they reduce the material costs of the formulation. Aromatic hydrocarbons are examples of diluents for nitrocellulose systems.

Application. Nitrocellulose finishes can be applied by almost all application techniques. Recoating times, which are totally dependent upon solvent loss, are typically in the region of 15 minutes to 2 hours. Three- or four-coat systems are normal, generally a specific primer or sealer followed by sufficient coats of a finish to produce the required film-build. For certain applications, however, such as the dip-coating of small timber items, toys, brush handles and the like, one-coat finishes can be formulated.

It is almost impossible to give representative formulations for the full range of nitrocellulose systems; in Tables 5.11 and 5.12 examples are given

Table 5.12 Clear Cellulose Finish

	Weight (%)
Cellulose nitrate, low-viscosity grade, damped 30% with alcohol	16·0
Short-oil-length coconut oil alkyd, 50% in xylene	10·0
Ester gum	15·0
Dibutyl phthalate	4·0
Methyl ethyl ketone	6·0
Methyl isobutyl ketone	16·0
n-Butanol	4·0
Isopropyl alcohol	4·0
Toluol	25·0

of a sealer for interior timbers and a clear lacquer topcoat, respectively. Surface defects in the final dried coat of the finish shown in Table 5.12, such as 'orange-peel' or small scratches, can be removed by the process known as pulling-over. This entails redissolving and smoothing the surface of the lacquer with a rag application of a pull-over solvent composition (which contains only small quantities of active and latent solvents but large amounts of diluents). When dried, the final defect-free finish is then burnished to develop the required gloss level.

5.5 Bibliography

Richardson, B.A., *Wood Preservation*. Longman: New York, 1978.

Newton, D.S. (Ed.), *Paint Technology Manual*, Part 4, *The Application of Surface Coatings*. Chapman and Hall: London, 1965.

Flick, E.W., *Water-based Paint Formulations*. Noyes Data Corporation: New Jersey, 1975.

Collier, J.W., *Wood Finishing*. Pergamon: Oxford, 1967.

Flick, E.W., *Solvent-based Paint Formulations*. Noyes Data Corporation: New Jersey, 1977.

CHAPTER 6

Formulation of Coatings for Masonry

6.1 Introduction

The main substrates considered in this section are concrete and cement-rendering, both of which are normally based on the inorganic binder, cement.

Most constructional cements are hydraulic, that is, they set in the presence of water, and by far the most widely used type is Portland cement. Portland cement is a finely-ground powder which, when mixed with water, sets and develops strength primarily because of the large content of anhydrous calcium silicates. These and other constituents of cement react chemically with water (by hydration and hydrolysis) and the final structure is generally considered to be a colloidal gel of high internal surface area.

Portland cement is typically prepared from limestone together with clay or shale. The raw materials are finely ground and heated to about 1500°C, which decomposes the calcium carbonate to yield the calcium oxide. A partially fused clinker is produced which is ground to produce the cement. Prior to this grinding, some calcium sulphate is generally added in order to regulate the setting time of the cement.

Cements with special properties are obtained by varying the ratio of the raw materials and by the incorporation of chemical additives at the clinker grinding stage.

Portland cement constitutes the binding agent in most constructional concrete and cement-renderings. It is used in admixture with sand and mineral aggregates, which modify the hardness and durability characteristics of the finished product and also reduce the shrinkage during setting to acceptable levels. Concrete, which occupies a dominant position in modern construction work, is a mixture of Portland cement, sand and graded aggregates, typically in the volume ratio 1:2:4. Cement-rendering is a relatively thin-section covering applied to building surfaces such as brickwork and blockwork to improve weather resistance. Neat cement–sand mixtures are used in certain situations, although hydrated lime is frequently added, typically in the ratio of 1:2:6 cement:lime:sand, to offset shrinkage and cracking.

Dry, mature and sound surfaces of this type present no difficulties in painting. However, when the surfaces are newly applied, or weathered to the extent that breakdown has occurred, they are among the most troublesome encountered.

The major source of painting problems on new cementitious surfaces arises as a result of residual moisture from the water addition required to effect curing of the cement, as well as the presence of alkaline constituents within the cement itself. The combination of these two factors can be particularly destructive to paints sensitive to alkali attack and therefore resistant types of coating have to be used under these conditions.

When they have dried out, the risk of alkaline attack from cementitious backings is reduced, although the danger of reactivation of the salts by subsequent wetting is still present. It is only after extensive periods of weathering that the alkali salts are rendered harmless by reaction with atmospheric carbon dioxide, and because of this the use of alkali-resistant coatings on concrete and cement-renderings is generally considered desirable.

Other problems experienced with these substrates include variable, sometimes excessive, porosity and surface friability and cracking. Both conditions are generally found on surfaces that have received prolonged weathering or have undergone severe conditions of exposure. Painting problems due to varied porosity across the surface are, however, also encountered on new, dry cementitious substrates.

6.2 Paint systems for cementitious surfaces

Although colour and surface appearance are usually the prime considerations when choosing paints, the characteristics of the substrate itself are important. Thus the moisture content, alkalinity and surface integrity of the substrate, as well as factors such as the service environment and exposure conditions, must also be taken into account.

Generally, the formulation of paints for cementitous surfaces assumes the presence of alkali salts and consequently some resistance to high pH conditions is built in at the design stage. This can be either by using alkali-resistant polymers and pigments or by using an alkali-resistant sealer (or primer) coat prior to the application of topcoats, particularly when the latter may have a lower tolerance to attack. Examples of resistant polymers include chlorinated rubber as well as a variety of vinyl resins, both solvent-soluble and emulsion types being used. Resistant pigments include titanium dioxide, chrome oxide, iron oxides, carbon black and the phthalocyanines.

The most important consideration when painting cementitious substrates, however, is the moisture content at the time of paint application. Excessive moisture affects the adhesion of most paints and can influence the ultimate performance of the applied system. Even in good drying conditions, many months are required to dry cementitious substrates. Rough estimates of one week of dry weather for every 5 mm thickness of wet construction are generally accepted as adequate for drying.

In the various sections of this chapter the commonly encountered conditions of cementitous substrates are categorised, and the types of coating that can be applied, and their main principles of formulation, are discussed under each heading.

6.3 Coatings for interior substrates

Concrete and cement-render are not frequently encountered in interior situations, the general practice being to apply gypsum (plaster) coats to these substrates. The formulation of paints for plaster is discussed in Chapter 7.

6.4 Coatings for exterior substrates

6.4.1 EXTERIOR SUBSTRATES: FRESH, WET

The painting of new concrete and cement-rendered surfaces is difficult. Cement paints are suitable for very early decoration and they can be applied within a few days of the substrate setting.

Cement paints are based largely on Portland cement with specific additives to control application, setting time and colour, and clearly the tolerance of these paints is due to the similarity between the paint and the substrate. A typical formulation of a white cement paint, showing approximate limits, is shown in Table 6.1. As can be seen, cement paints are supplied in powder form and are mixed with water to a suitable consistency

Table 6.1 Cement Paint, White

	Weight (%)
White Portland cement	64–94·5
Hydrated lime	25–0
Calcium or aluminium stearate	1–0·5
Calcium chloride	5–2·5
Pigment, titanium dioxide	5–2·5

prior to use. Application is generally two coats by brush, allowing 16–24 hours between coats.

The level of pigmentation is fairly critical and generally the quoted 5 per cent figure is not exceeded. Additions of pigment above this level can result in reduced film integrity of the resultant paint. However, addition rates of the order of 5 per cent and less are adequate to produce a wide range of colours and, where darker shades are required, the use of ordinary Portland cement which is grey in colour is normal. The types of pigment that can be used are restricted to those of high tinting strength (because of the low addition rates), as well as to those with high alkali resistance. Inorganic pigments are generally favoured.

The other additives in the formulation modify the application rate and film properties. The calcium or aluminium stearate addition improves the application characteristics of the paint and also upgrades the water resistance of the dried film. Calcium chloride, which accelerates the setting of Portland cement, is added to regulate the drying time of the paint. Hydrated lime is sometimes added to cement paints where it functions as an extender for the cement constituent and reduces cracking tendencies.

6.4.2 EXTERIOR SUBSTRATES: FRESH, DAMP

Other paint types can safely be applied to new cementitious surfaces which have had about 4 weeks' drying. After this period, suitable surfaces would not be expected to show any signs of surface moisture. However, in-depth moisture contents would be appreciable, especially where thick sections are employed, and for this reason relatively porous coatings are preferred. This ensures that the moisture released during the prolonged drying-out period does not disrupt the applied coatings.

Emulsion paints. The main group of coatings suited to application under these exacting conditions is based on emulsion polymers, since they tend to have a greater permeability in comparison with similar coatings based on solvent-soluble resins.

The principal ingredients of an emulsion paint comprise the pigmentation and the binder, the latter being a polymer in the form of an aqueous dispersion. The pigments and extenders used are broadly similar to those used in solvent-soluble types of paint.

The pigments and extenders are dispersed into water which functions as the volatile 'carrier' component of the paint. The water also contains small amounts of pigment-dispersing agents as well as thickening agents which control the ultimate viscosity of the paint. Typically, the dispersing agents are sequestering or surface-active agents such as sodium hexametaphosphate (about 0·5 per cent on weight of pigment) and the thickening agents are cellulose derivatives (0·2–1·0 per cent on weight of paint) or, less frequently, for masonry coatings intended for this specific application, alkali-soluble polyacrylates.

The pigment paste is then blended with the emulsion polymer and certain minor additives essential to the stability of the liquid paint and to facilitate film formation are incorporated. These additives include anti-foam agents such as specific organo-silicon compounds, fatty acid esters and fatty acid metal soaps, added at less than 0·1 per cent weight, which displace the air foam formed in the paint during manufacture and characteristic of most types of emulsion paint.

Biocides (or bactericides) and in certain situations fungicides are also added, biocides to control bacterial degradation of the liquid paint during storage, and fungicides to eliminate mould growth on the coating during service. Addition levels of these compounds vary with the type of material being used; typically, biocides are added at less than 0·5 per cent by weight of paint, whereas fungicides are added at varying levels up to about 2 per cent, depending on the efficacy of the compound as well as the desired service life.

A wide range of complex organic compounds have bactericidal or fungicidal action in emulsion paints, although most tend to be specific. Examples of biocides used are chlorinated diphenyls, quaternary ammonium compounds and fluorine-containing sulphamides. Typical fungicides that find use are chlorinated phenols, as well as dithiocarbamates and isothiazolin compounds.

Emulsion polymer. The emulsion polymer used in these paints must impart the necessary attributes of adhesion, durability, water resistance and chemical resistance to the coating. Several types have found widespread use in masonry coatings: examples include pure acrylic polymers, styrene-modified acrylics , polyvinyl acetate (p.v.a.) modified with acrylic and p.v.a. modified with other co-monomers such as versatic acid and ethylene.

Emulsion polymers differ from solvent-soluble polymers in that they are not solutions but are in the form of a fine particle-size (0·3–3·0 μm) suspension in water. The mechanism of film formation of these polymer emulsions and of the paints derived from them is complex and involves loss of the water phase into the substrate and/or the atmosphere. At the same time, the polymer particles flow together or coalesce, leading to film integration. Effective coalescence of emulsion particles, however, will not occur below a certain critical temperature which is a characteristic of the particular emulsion system. In practice, if this temperature is too high, it can be depressed by including in the emulsion paint formulation a small quantity of a coalescing aid, which is a specific high-boiling-point organic solvent, typically the acetate of ethylene glycol monobutyl ether, aromatic glycol ethers or complex ester alcohols or, less commonly, propylene glycol.

Pigmentation. Emulsion paints formulated for application to masonry which has not fully dried are generally of moderately high pigment volume

Table 6.2 Exterior Emulsion, White

	Weight (%)
Titanium dioxide, rutile	25·0
Talc	11·0
2% Solution of cellulose thickener	16·0
10% Solution of sodium hexametaphosphate pigment dispersant	1·5
Defoamer	0·1
Bactericide	0·3
Water	9·4
Vinyl acetate–acrylate copolymer emulsion, 54% solids	34·2
High-boiling ether coalescing solvent	2·0

concentration, although the level adopted has to be consistent with adequate film durability. Pigment volume concentration levels of 35–45 per cent have been successfully used in this application and this range permits subsequent drying-out of the substrate without film disruption.

Typical formulations for matt finish white exterior quality emulsion paints of about 36 per cent and 42 per cent P.V.C., respectively, are shown in Tables 6.2 and 6.3. These coatings would be brush-applied in two coats directly onto the substrate.

Colours can be produced using suitably alkali-resistant pigments and/or pigment paste tinters. Where deep colours are required, especially those produced from organic pigments, a tint base formula where the white pigment is totally excluded or reduced to a minimum is normally used, the P.V.C. being maintained by means of a suitable extender combination. Deep-tint-base formulations are discussed further in the section on paints for plaster.

Table 6.3 Exterior Emulsion, White

	Weight (%)
Titanium dioxide, rutile	24·2
Talc	15·8
2·5% Solution of cellulose thickener	7·2
10% Solution of polyphosphate dispersant	1·2
Defoamer	0·1
Water	7·8
Acrylic emulsion, 50% solids	38·8
Propylene glycol	2·8
Ethylene glycol	2·0
Bactericide	0·1

Textured finishes

Textured emulsion finishes are very popular and widely used on cementitious substrates. These are generally capable of providing higher film-build compared with the standard flat finish types of emulsion paint discussed earlier. To compensate for the lower permeability of these higher film-build paints, the P.V.C. of the formulation is usually increased. It is normally found, however, that despite the high P.V.C. levels, textured paints tend to be very durable on masonry surfaces.

The commonly-used texturing agents vary in particle size from about 0·2 to 2·0 mm, and sometimes even larger. In general, refined sands or coarsely ground paint-extender materials are used to produce the finer-textured (or sand-textured) paints, whereas ground flints, various silicates or particulate carbonates are used to produce the coarser-textured (or aggregated) paints.

At the formulation stage, close attention has to be paid to the grading of the texturing agent to ensure that it is consistent with the mode of application as well as capable of producing the correct packed appearance in the dried film.

Sand-textured emulsion paints are manufactured on similar lines to conventional types, except that the texturing agent is added after pigment dispersion has been achieved, but prior to any necessary colour-matching adjustments. The formulation of sand-textured paints differs somewhat from that of flat-finish paints and, in particular, lower levels of pigment can be achieved with concomitant higher loadings of the extender texturing agent. This is not at the expense of film opacity, since the higher film-build of these types of finish compensates for any loss of opacity. A typical formula is shown in Table 6.4, and this would be applied in one or two coats by brush or spray. The pigment volume concentration of this formulation is about 48 per

Table 6.4 Sand-textured Emulsion Paint, White

	Weight (%)
Titanium dioxide, rutile	14·5
Whiting	7·0
Talc	15·0
Water	17·8
Sodium hexametaphosphate	0·15
Bactericide	0·1
Cellulose thickener	0·25
Styrene acrylic emulsion, 50% solids	25·0
Coalescing solvent*	1·2
Washed sand, graded 1000–355 μm	13·0
Washed sand, graded 355–255 μm	6·0

*Ethylene glycol monobutyl ether.

cent. The validity of applying the P.V.C. concept to textured paints has not yet been established; nevertheless, it is assumed here that the addition of particulate sands or aggregates does not significantly affect the P.V.C. of the pigment–extender–polymer combination used in a particular formulation.

Aggregate-textured paints are formulated as in Table 6.5. This coating would be applied by plasterer's trowel or specially designed spray gun. The colouring effect is obtained by the use of self-coloured mineral aggregates, marbles, etc., or by means of synthetically coated flint aggregates, typically with polyurethane-based paints. Very small amounts of pigment powder can be added during manufacture to control the final shade of the coating. Manufacture is by simple blending, no grinding or attrition being required.

Table 6.5 Aggregate-textured Paint

	Weight (%)
Precoated flint aggregate, 0·15 mm	76·0
Talc	4·0
Water	2·4
Cellulose thickener	0·05
Bactericide	0·05
Wetting agent (surface-active agent)	0·1
High-boiling ether coalescing solvent	0·4
Poly(vinyl acetate)–acrylic copolymer, 55% solids	17·0

A high aggregate-to-binder solids ratio (about 8:1 in the example shown), needs to be at an optimum so that the polymer just coats and binds the particles of aggregate and affords suitable wet adhesion. A surfeit of polymer tends to form a film which partially masks the topography of the dried paint. The talc used in the formulation improves the water resistance of the dried film. The wetting agent is an important component which is added primarily to improve the application characteristics when trowel-applied.

6.4.3 EXTERIOR SUBSTRATES: FRESH, DRY

At some stage after construction a state will be achieved when there is no excess moisture retained within the cementitious substrate. This interval will be dependent upon the thickness of the section as well as the severity of its exposure, and might be in the region of 1–12 months.

At this stage there is virtually no restriction to the coating types than can be applied, although precautions to guard against accidental wetting and reactivation of alkali salts include incorporation of alkali resistance into applied systems. Where systems of low inherent alkali resistance such as vegetable oil-modified alkyds are to be applied, a resistant sealer or primer coat should be provided. These sealers also have an important application in consolidating the surface of weathered cementitious substrates and are discussed in the following section.

Three main classes of paint are applied to dry cementitous surfaces, namely coatings based on chlorinated rubber and on vinyl copolymers, which are generally applied where both high durability and water impermeability are required, and emulsion paints with or without texture; the latter have already been described. Other paint types may also be applied on epoxy, polyurethane and alkyd bases, although their use is possibly not widespread compared with the above-mentioned classes. The formulation principles of these types of coating have been described in those sections where they are used most frequently, and accordingly are not detailed here. In general, however, the use of these polymers presents no special problems in paints for masonry, and they are normally all formulated as low-build systems with a semi-gloss or matt finish.

Of the coatings that are applied to masonry, it is probable that the emulsion-based systems are the most frequently encountered in general-purpose use. The chlorinated rubber and the vinyl systems find their main application in specially demanding situations where high durability and exceptional resistance to water and pollutants are required.

The formulation of textured emulsion paints for fresh, dry cementitious substrates is identical to that for coatings used on damper substrates. With the untextured emulsion paints, however, lower pigment volume concentration levels, about 30–35 per cent, can be adopted since high permeability to moisture is not required. This ability to decrease the system P.V.C. for dry cementitious substrates enables better durability characteristics to be achieved. A formulation of a white exterior emulsion paint of about 28 per cent P.V.C. is given in Table 6.6.

Table 6.6 Exterior Emulsion, White

	Weight (%)
Titanium dioxide, rutile	22·6
Talc	6·8
2% Solution of cellulose thickener	14·0
Tripolyphosphate pigment dispersant, 10% solution	0·6
Defoamer	0·2
Water	13·0
Poly(vinyl acetate)–acrylic copolymer emulsion, 55% solids	39·7
High-boiling ether solvent	1·0
Propylene glycol coalescing solvent and freeze–thaw stabiliser	2·0
Biocide	0·1

Table 6.7 Chlorinated Rubber Masonry Paint, White

	Weight (%)
Titanium dioxide, rutile	15·0
China clay	16·2
Chlorinated rubber, medium-viscosity grade	14·0
Chlorinated paraffin	9·5
Clay gellant	0·5
Industrial alcohol	0·2
Trimethylbenzene	33·5
White spirit	11·1

Chlorinated rubber-based systems

Paint systems based on chlorinated rubber give excellent performance on cementitious surfaces, the inherent alkali-resistance of the polymer making them ideally suited for this class of substrate. However, the characteristic low water-permeability of these coatings normally limits their use on cement render and concrete that contain large amounts of retained moisture, although in contrast the use of these coatings as primers on damp internal plaster surfaces is common.

Both low-build (two-coat, brush-applied) and high-build (single-coat, airless-spray-applied) variants of chlorinated rubber-based masonry paints are used. For cementitious surfaces, low gloss paints are generally preferred since this tends to mask minor topographic defects in the surface.

Both inert as well as resinous plasticisers are used in chlorinated rubber masonry coatings. Tricresyl phosphate, trichloroethyl phosphate and, expecially, chlorinated hydrocarbons, all at addition levels of plasticiser to chlorinated rubber of $<1{\cdot}0{:}1{\cdot}0$, are representative of the inert plasticisers, while varnishes with a high degree of alkali resistance (for example, tung oil–phenolic) are typical of the resinous plasticisers. The use of vegetable oil-modified alkyds as polymer plasticisers for chlorinated rubber is possible, although at the expense of some water resistance and alkali resistance. Resin modification does result in films with a lower tendency to embrittle during weathering compared with the inert plasticisers; addition levels of 1:1 to 4:1 resinous plasticiser : chlorinated rubber are usual.

Low-build coatings. The formulation given in Table 6.7 is an example of a low-build chlorinated rubber masonry paint plasticised with chlorinated paraffin, at a P.V.C. of about 40 per cent. Application would be by brush and two coats would be required, the first being diluted with some 10–20 per cent solvent on porous surfaces. The pigment volume concentration and plasticiser:chlorinated rubber ratio adopted here are typical for this application, although this P.V.C. is the upper limit for this particular

Table 6.8 Swimming Pool Coating, White

	Weight (%)
Titanium dioxide, rutile	20·0
Chlorinated rubber, medium-viscosity grade	18·0
Chlorinated paraffin	12·0
Clay gellant	0·5
Industrial alcohol	0·2
Trimethylbenzene	36·9
White spirit	12·3

combination of plasticiser and polymer. Adhesion may suffer at higher P.V.C.s. There is, however, some evidence to suggest that natural carbonate extenders confer superior resistance to freeze–thaw conditions and may be preferable to china clay on substrates of high moisture permeability, in particular brickwork and cement-render.

Apart from their use on cementitious substrates exposed to atmospheric weathering, these types of chlorinated rubber paint are used to coat continuously immersed surfaces such as swimming pool interiors. For the majority of pools, the best results are obtained using a lower P.V.C. coating than that shown in Table 6.7, levels of about 20 per cent being typical (Table 6.8). In certain instances, for example where the waterproof membrane or liner has been omitted or the surrounding ground water table is high, a higher P.V.C. paint would be preferable. The greater permeability reduces the tendency of the coating to blister during service.

An example of a chlorinated rubber–long-oil-length alkyd paint is shown in the section on coatings for structural steel. The main formulation problem with such coating systems is one of compatibility of the polymers used, although, in general, long-oil-length varnishes and alkyd resins are compatible. Chlorinated rubber, however, only has limited compatibility with medium- or short-oil-length materials, but compatibility does increase on decrease in the molecular weight of the chlorinated rubber. Paints based on chlorinated rubber–resinous plasticiser are always formulated to produce low-build films, since the oxidative drying process of the alkyd or varnish modifier would not be effective in very thick films.

High-build coatings for masonry can be formulated using chlorinated rubber–inert plasticiser combinations, and these are capable of establishing highly protective films with a one-coat application.

Where application is to be by airless spray, low-viscosity grades of chlorinated rubber are preferred with a solvent combination of xylene and 2-ethoxy acetate, and hydrogenated castor oil as gellant. A small amount of lamellar extender such as micronised mica can be included and this is beneficial in reducing excessive binder absorption, especially on substrates of high or variable porosity.

Pigment volume concentrations in the region of 35–40 per cent are typical for high-build systems. Table 6.9 shows a formulation of a flat finish for cementitious substrates capable of establishing dry film thicknesses of about 100 μm with a one-coat airless-spray application. The P.V.C. is about 40 per cent. The pigmentation of any of the types of chlorinated rubber-based

Table 6.9 High-build Chlorinated Rubber, White

	Weight (%)
Titanium dioxide, rutile	18·3
Barytes	13·0
Calcium carbonate	9·4
Mica	1·0
Chlorinated rubber, low-viscosity grade	14·3
Chlorinated paraffin	7·7
Hydrogenated castor oil	0·8
Xylene	28·4
2-Ethoxy acetate	7·1

paints described above presents no major problem, although certain pigment types have to be avoided. These include Prussian blue, ultramarine blue, Brunswick green, lead chromates and organic pigments of the toner class. For application to cementitious substrates, however, the pigments adopted would also need to be alkali-resistant in their own right.

Vinyl polymer-based systems

Systems based on various solvent-soluble vinyl polymers, or solution vinyl polymers, are also used on cementitious substrates. These materials possess some similarity to the chlorinated-rubber-based systems, in particular very high durability, but have somewhat lower resistance to chemicals and water.

Coatings based on the solution vinyl resins are generally of low film-build and are designed for brush application in two coats, although high-build variants for one-coat application by airless spray can be formulated.

Many types of vinyl resin are used as the basis for coatings for cementitious substrates, typically the copolymers of polyvinyl acetate and polyvinyl chloride, as well as styrene butadiene, vinyl toluene butadiene, styrene acrylate and vinyl toluene acrylate. These latter types have the advantage, compared with the p.v.a.–p.v.c. copolymers, of being soluble in aromatic–aliphatic solvent blends with the attendant benefits of using such blends. The copolymers of p.v.a.–p.v.c. are normally used as the basis of the high-build vinyl systems, and the main area of application for this type of coating is in anti-corrosive finishing systems for steelwork. Low-build coatings for plastics also constitute an important use for the p.v.a.-p.v.c. copolymers.

The vinyl masonry coatings are generally based on styrene butadiene, vinyl toluene butadiene, styrene acrylate or vinyl toluene acrylate; all of these polymers are similar in terms of durability, pigment binding power and adhesion. However, the colour stability of the acrylic copolymers is superior

Table 6.10 Vinyl Masonry Paint, White

	Weight (%)
Titanium dioxide, rutile	17·0
Calcium carbonate	17·0
Mica	6·6
Diatomaceous silica	5·7
Vinyl toluene acrylate, high-viscosity grade	8·4
Modified acrylate gellant	1·9
Chlorinated paraffin	3·9
White spirit	33·7
High-flash naphtha	5·8

to that of the butadiene copolymers, which are therefore preferred in white and pastel-shade formulations. As with chlorinated rubber-based paints, some plasticiser is added to the formulations; the chlorinated paraffins are preferred. Ester plasticisers, such as dibutyl phthalate and dioctyl phthalate, can be used but they are not alkali-resistant and moreover they can migrate from the film. Vegetable oils and varnishes are of only limited compatibility with solution vinyl resins.

The high pigment-binding capability of the solution vinyl resins enables higher pigment volume concentrations to be adopted compared with the chlorinated rubber systems; levels of about 58 per cent for the butadiene copolymers and about 52 per cent for the acrylic copolymers are recommended for masonry coatings. The formulation shown in Table 6.10 is for a white masonry coating designed for brush application in two coats. The resin used here is a high-viscosity grade of vinyl toluene acrylate thickened with a specific modified vinyl toluene acrylate resin. However, alternative thickening agents such as hydrogenated castor oil or amine-treated clays can also be used successfully.

The pigment volume concentration of this formulation is about 53 per cent and the pigment–extender combination shown has been found to be optimum for this type of coating. There are no major limitations, however, on the types of pigment or extender that can be used with solution vinyl resins.

Table 6.11 Clear Sealer

	Weight (%)
Hydrocarbon resin	35·0
White spirit	65·0

6.4.4 EXTERIOR SUBSTRATES: WEATHERED

Weathered cementitious substrates can present problems for the paint formulator because of the nature of the resultant surface. In particular, cracks can open as a consequence of the shrinkage process of cement but, more importantly, the outermost layer of the surface can become weakened and friable. This presents a poor base for paints and precautions need to be taken. Ideally, all weakened surface material should be removed by abrasion prior to painting operations. Often this is not possible, particularly on larger jobs, because of cost; in such cases, all that may be possible is a more or less superficial wiping-down with either a stiff broom or a wire brush.

Sealers

In this situation some form of sealer coat is advisable, since the low viscosity characteristics of these materials, and hence their penetrative ability, entail a concomitant binding or consolidation of the substrate surface. Sealer coats also have applications where the main or subsidiary requirement is to even-out variable surface porosity, a frequent problem with cementitious surfaces.

Sealer coatings are generally based on solvent-soluble resin systems since these have a greater ability to penetrate compared with the emulsion-based systems; pigmentation is sometimes dispensed with altogether. Frequently, however, more paint-like formulations are used, these functioning as combined primer–sealer coats: that is, they are sufficiently fluid to penetrate and consolidate the surface, yet their solids volume is such that film-build is significant. Pigment content, however, needs to be kept reasonably low since too high a level would interfere with penetration. Levels are normally sufficient to impart only moderate opacity consistent with giving a visual indication of coating uniformity during application.

The formulations shown in Tables 6.11 and 6.12 give examples of each type of sealer, alkali resistance again being a requirement. Gellants or thickening agents should either be omitted or only be used at very low levels in these types of formulation. Too high a proportion will have adverse effects on penetration and hence surface binding.

Table 6.12 Primer–Sealer

	Weight (%)
Titanium dioxide, rutile	10·0
Medium-oil-length ester gum–tung oil varnish, 60% in white spirit	20·0
Chlorinated rubber, medium-viscosity grade, 25% solution in 3:1 xylene: white spirit	40·0
Xylene	25·0
White spirit	5·0

The very simple sealer formulation in Table 6.11 can be improved upon by the substitution of about 10–20 per cent of the hydrocarbon resin by a low-viscosity solvent-soluble vinyl resin such as vinyl toluene acrylate, which will improve the binding capability of the sealer.

These types of sealer or primer–sealer are widely used as pretreatments on cementitious substrates prior to application of the finishing coats. On mature surfaces, where little if any residual uncombined water is present in depth, there is virtually no restriction on the formulation of the finishing system, provided that it has the necessary durability and performance attributes.

Finishes

Any of the coatings described earlier can be used on weathered surfaces, often in conjunction with the general-purpose sealer types shown in Tables 6.11 and 6.12. Where the more sophisticated polymer types are used as the basis for the finishing system, i.e. chlorinated rubbers, solution vinyls, epoxies and polyurethanes, specifically designed sealers based on a similar resin base are often employed. This is a safeguard against possible incompatibility.

When cement paints are applied to mature cementitious surfaces, no sealer coat is necessary. It is desirable, however, to pre-soak the substrate with water before painting operations begin; this ensures that the water required to cure the cement is not abstracted from the coating by the substrate as it is applied.

Sealers and primer–sealers are therefore most frequently used in combination with both textured and untextured emulsion paint systems, while alkyd-based finishes can also be successfully applied to mature cement surfaces. The necessary protection against the risk of alkali attack on the susceptible alkyd resin component is offset by the inherent resistance of the sealer. A formulation for an alkyd paint suitable for cementitious substrates is shown in Table 6.13; in common with other finishes for masonry, it has a low sheen. The pigment volume concentration here is about 34 per cent and the paint would be applied in two coats over a sealed surface.

Table 6.13 Alkyd Masonry Paint, White

	Weight (%)
Titanium dioxide, rutile	31·4
Micronised barytes	10·5
Medium–long-oil-length semi-drying-oil alkyd, 50% in white spirit	41·9
White spirit	14·6
Anti-skinning agent, methyl ethyl ketoxime, 10% solution in white spirit	1·0
24% Lead naphthenate	0·43
6% Cobalt naphthenate	0·17

Apart from the higher P.V.C. adopted, the formulation principles of these masonry paints follow similar lines to those of the alkyd-based systems applied to other building substrates. General formulation considerations have been discussed on page 108.

Pigment volume concentrations for alkyd-based finishes for cementitious surfaces are typically in the region of 30–50 per cent, although higher levels can be used successfully. Unlike the alkyd systems applied, for example, to timbers, the system on masonry is built up by a coat-on-coat application of the same product, rather than by the use of an undercoat and topcoat of differing compositions. The higher P.V.C. level adopted, apart from reducing the gloss level generally desired on this type of substrate, also provides the film with the ability to allow the transmission of water, an obvious advantage on cementitious substrates.

Although there are no major limitations on the pigments and extenders that can be used with alkyd paints, alkali resistance is an advantage but not a prime requirement on mature surfaces. However, in order to maintain adequate opacity in white and light-tint shades, at least 50 per cent of the pigment–extender content should be titanium dioxide.

6.5 Bibliography

Flick, E.W., *Water-based Paint Formulations*. Noyes Data Corporation: New Jersey, 1975.

Flick, E.W., *Solvent-based Paint Formulations*. Noyes Data Corporation: New Jersey, 1977.

Newton, D.S. (Ed.), *Paint Technology Manual*, Part 4, *The Application of Surface Coatings*. Chapman and Hall: London, 1965.

CHAPTER 7

Miscellaneous Substrates

7.1 Introduction

There are obviously surfaces and exposure conditions other than those covered in Chapter 4, 5 and 6 for which the paint technologist may be required to formulate coatings. Such formulations may, in the same way, be based on consideration of the properties of the substrate and the conditions of use, followed by appropriate raw material selection. In this chapter, paints for application to gypsum plaster, to non-ferrous metals and to plastics are discussed.

7.2 Formulation of coatings for plaster

7.2.1 INTRODUCTION

Plaster as a substrate is encountered only in interior situations. The various difficulties it presents as a substrate for paint depend on factors such as its moisture content and chemical composition.

Modern plasters are all based on gypsum, which is the common name for the dihydrate of calcium sulphate, $CaSO_4.2H_2O$. During manufacture, gypsum is heated to dehydrate the combined water and yield the hemihydrate form of plaster, which approximates to the composition $2CaSO_4.H_2O$. During the manufacturing process, specific additives are incorporated to modify and control the setting characteristics of the resultant plaster.

In use, water is added to the plaster and the paste is applied by trowelling to the backing, which may be other coats of plaster or sand–cement rendering. Setting of the plaster involves a hydration process which reverts the paste back to its dihydrate form, the final stage obtained in the overall setting process.

There are four basic types of plaster. The first is plaster of Paris which is a form of hemihydrate without set-retarder; this material is now rarely encountered in anything but restoration work. The second type are retarded hemi-hydrate plasters, which are a form of hemihydrate containing added set-retarders. Anhydrous gypsum plaster is a third form of plaster and consists essentially of anhydrous calcium sulphate. Lastly, there is Keene's plaster, also an anhydrous plaster but one which is characterised by being more easily brought to a smooth finish in a continuously plastered surface and, furthermore, which has a more gradual setting rate than the anhydrous plasters.

The retarded hemi-hydrate plasters are also obtainable with added lightweight materials such as perlite or vermiculite. These additions are made for situations demanding high fire resistance and good thermal insulation. Most plasters are initially slightly alkaline, although this quickly reduces to neutrality on ageing. Lime may be added both during manufacture and by the applicator to improve workability during trowelling.

7.2.2 PLASTER AS A PAINT SUBSTRATE

The adhesion of paints to plaster is good even on the denser, anhydrous calcium sulphate types, although highly trowelled surfaces can present problems. The addition of lime to plaster can also be a problem, especially if saponifiable paints are being used and where dampness is present. However, nowadays the majority of plaster is painted with emulsion-based formulations, which are invariably highly alkali-resistant, and the lime addition rarely presents problems in practice. Alkyd-based paints can be used successfully in dry interior situations.

Greater problems can arise from the variable porosity caused by uneven trowelling during plastering, since this results in uneven paint uptake. The problem is especially noticeable with emulsion paints, the finished surface appearing patchy in its sheen and, possibly, even its colour. The remedy is either to use a penetrating sealer (see page 158) or to dilute the first coat of emulsion prior to application.

Efflorescence

A further problem associated with plaster is salt damage to dried films—the phenomenon known as 'efflorescence'. Efflorescence damage to applied coatings is associated with the presence of water-soluble inorganic salts within the substrate. With gypsum plasters, the salt is essentially sodium sulphate and arises either as an impurity in the plaster itself or as a result of migration from cementitious or brickwork backings. The movement of these salts during the drying-out phase of the plaster can cause severe disruption of applied coatings, and is a common cause of paint failure in new buildings.

Resistance to efflorescence can only be completely developed by allowing the plaster to dry out thoroughly before painting. Where this is not possible, paint coatings of low water permeability, which also provide a uniform surface film, will exhibit acceptable resistance. Coatings that are over-pigmented or which penetrate excessively are not resistant, nor are most emulsion paints of the type used on plaster, since they tend to be too permeable.

However, the prior use of a chlorinated rubber sealer of about 15 per cent pigment volume concentration (see page 159) produces good results, although clearly the sealer formulation itself must be resistant to efflorescence. The concept of applying highly pigmented emulsions to plaster in order to facilitate salt movement towards the surface, where it can be brushed off without damage to the film itself, is now no longer held to be tenable.

7.2.3 PAINTS FOR PLASTER

The most widely used paints for plaster are undoubtedly emulsion paints. Types similar to those described on pages 148–150 can be used. However, the less demanding service conditions mean that high pigment volume concentrations, up to about 75 per cent, can safely be adopted. These paints are low-cost materials and clearly are inadequate for many situations—for example, the wet abrasion characteristics would not be expected to be very good. Nevertheless, for general-purpose use in non-critical areas, their performance and appearance are acceptable. Table 7.1 shows an example of a white interior-quality emulsion paint of 72 per cent pigment volume concentration. This would be brush- or roller-applied in two coats.

Table 7.1 High-P.V.C. Interior Emulsion, White

	Weight (%)
Titanium dioxide, rutile	8·0
China clay	19·5
Whiting	16·2
Poly(vinyl acetate) acrylate copolymer emulsion, 54% solids	11·6
2% solution of cellulose thickener, low-viscosity grade	25·0
5% solution sodium hexametaphosphate	1·0
Bactericide	0·2
Water	18·5

The content of the opacifying titanium dioxide in these formulations can be reduced to very low levels, because of the dry hiding ability of extenders when incorporated into paints at very high loadings.

The technique of formulating these high P.V.C. emulsion paints is to work at low total solids, and to maximise the effectiveness of the water in the composition by using a low-viscosity grade of cellulose thickener, avoiding extenders with very high water absorptions. Coalescing solvents are not normally necessary in such formulations, although for winter use, especially in unheated buildings, some addition is advisable.

Generally, however, emulsion paints of lower pigment volume concentrations, about 40–50 per cent, are better suited for use in areas such as kitchens or other rooms where the paintwork might require to be washed.

Tint-base formulations

For ease of interpretation and cross-relation, the formulations described in this book are largely for white paints. In most practical situations, of course, a range of colours is required. Where relatively large additions of organic pigments are to be added to produce deep shades, these are usually added as tinters, that is, as stable dispersions in a compatible base; the so-called 'tint-base' formulations are commonly used.

Tint bases can be prepared in bulk for use as required in order to produce special shades to order. Two types are normally used, a deep tint base and a light tint base for the intermediate colours. Pigment volume concentrations similar to the white and full-shade examples of the appropriate product are maintained by the use of extender materials. An example of a deep tint emulsion base of about 35 per cent P.V.C. is shown in Table 7.2.

The intermediate tint base formulation would normally contain a proportion of titanium dioxide. The advantage of the tint-base approach is

Table 7.2 Deep Tint-base Emulsion

	Weight (%)
China clay	10·0
Whiting	21·6
Poly(vinyl acetate) ethylene copolymer emulsion, 55% solids	41·7
2% Solution cellulosic thickener	9·0
5% Solution sodium hexametaphosphate	2·5
Defoamer	0·1
Bactericide	0·1
Water	13·0
High-boiling ether coalescing solvent	2·0

that relatively consistent pigment volume concentrations can be maintained across a product range of colours and, furthermore, broadly similar opacity levels and film thicknesses can be obtained. Where deep shades are required that can be obtained using the relatively cheaper inorganic pigments, it is normal specifically to formulate a product using the particular pigment or combinations of pigments; final adjustments to shade are performed using organic tinters. The tint-base formulation is used mainly for those paints in which relatively large amounts of organic tinters are required. Clearly, some pastel shades can readily be achieved simply by tinting a white paint.

There are of course exceptions to the above. In particular, where a high gloss level is required in a finishing paint, extenders have to be excluded from the formulation. This can result in certain colours having significantly lower pigment volume concentrations; for cost reasons, pigments are only added in sufficient quantity to give the required opacity. For example, a white alkyd gloss would typically have a pigment volume concentration of 15 per cent, whereas its equivalent black shade, based on carbon black, might have a P.V.C. < 5 per cent. With many formulations, obtaining acceptable opacity within the cost limitations of the market becomes a major problem. It is clearly of great importance, however, that the paint technologist recognises the significance of composition variations across a product range in terms of subsequent coating performance.

Semi-gloss emulsions

Semi-gloss emulsions are also used on plaster to an increasing extent, largely because of their aesthetic appeal in many interior situations. Various resins can be used in these formulations, such as poly(vinyl acetate) copolymers, as well as pure acrylic and acrylic copolymer emulsions. Where high gloss level is required in a product for interior and exterior use, the latter classes of polymers are to be preferred. Pigment volume concentrations of about 15–20 per cent are used, and with the semi-gloss formulations higher proportions of extenders as well as somewhat higher P.V.C. levels can be used. The flow characteristics of the applied film, however, become of increasing importance with these higher gloss types of emulsion paint. Generally, the standard cellulose thickeners, such as sodium carboxymethyl cellulose or ethyl hydroxy ethyl cellulose, are not capable of giving the required rheological characteristics. In these types of formulation, the alkali-soluble acrylic thickeners described on page 136 can be used to obtain better flow properties.

Flat and eggshell sheen paints

Flat and eggshell sheen alkyd-based paints may also be used on interior plasters. These are similar in composition to the types used on dry cementitious surfaces (page 160), except that higher pigment volume

Table 7.3 Interior Flat Alkyd Finish, White

	Weight (%)
Titanium dioxide, rutile	17·0
Whiting	38·4
Silica	2·5
Medium–long-oil-length semi-drying oil-modified thixotropic alkyd, 50% in white spirit	29·0
White spirit	11·7
Methyl ethyl ketoxime, anti-skin agent, 10% solution in white spirit	1·0
24% Lead naphthenate	0·28
6% Cobalt naphthenate	0·12

concentrations can be adopted. The formulation shown in Table 7.3 for a flat white finish has a pigment volume concentration of about 55 per cent and demonstrates the use of a thixotropic alkyd resin. This would be brush-applied in two coats.

The thixotropic alkyds impart a gelled structure to paints without the use of additional treated clay gellants or alternatives. They are especially suited for applications where good flow is not a primary requirement but where ease of brushing is important. Yellowing during service can be a problem with the thixotropic alkyd paints. This is due to the presence of polyamide materials, which can have poor colour stability. However, in interior situations these paints are generally satisfactory.

More conventional alkyd binders can be used as an alternative in these types of paint, especially where better flow properties are required, i.e. in eggshell or semi-gloss finishes. The use of medium-oil-length alkyds is satisfactory for interior use on plaster, although longer-oil-lengths alkyds can be used as the basis of interior–exterior flat or low-sheen paints.

7.3 Formulation of coatings for non-ferrous metals

7.3.1 INTRODUCTION

Several types of non-ferrous metal require coating, including aluminium, zinc, zinc-coated steel and galvanised iron. As with other metal substrates, the surface has to be adequately prepared prior to receiving paint coatings although, unlike mild steel, the presence of oxide layers on the surface is not necessarily detrimental to performance provided that atmospheric contaminant salts have not been deposited within their structure. With galvanised steel in particular, preweathering for a period of several months followed by washing with clean water is thought to be beneficial to subsequent coating adhesion. Aluminium metal, in its finished state, has an adherent oxide layer that is protective in its own right, particularly when formed electrochemically as in anodising.

A detailed study of the surface preparation of non-ferrous metals is outside the scope of this book. It is considered at greater length in certain of the texts listed in the bibliography.

Basically, however, the preparation of non-ferrous metals requires the removal of greases and oils, frequently applied as temporary protectives after preparation, as well as dirt. This degreasing is performed in a similar manner to that adopted for mild steel, for example, by vapour degreasing or by wiping with solvent swabs under site conditions.

If the metal has been subjected to exterior exposure, loose oxide layers, dirt and salt deposits need to be removed, and this is accomplished by relatively simple techniques such as wire-brushing or sanding. Indeed, the subsequent performance of the component could be seriously affected if the natural protective oxide layers, or deposited zinc metal layers on steel, were removed by over-vigorous cleaning procedures such as shot-blasting or sand-blasting.

For both aluminium and zinc, factory-applied chemical pretreatments have also been developed similar to the phosphating processes applied to steel.

7.3.2 PAINTS FOR ALUMINIUM

In common with most other non-ferrous metals, aluminium generally has a greater corrosion resistance than steel and, accordingly, the choice of surface coatings is less critical. In many exterior situations the conventional alkyd-based undercoat and gloss topcoat combinations are adequate. Where more aggressive environmental situations are encountered, around chemical plant for example, the low build chlorinated rubber or epoxy-based paints can be used. The formulation of these topcoats follows a similar pattern to those formulated for steel.

The most widely used primer type for aluminium is the etch primer as described for use on ferrous metals (page 99). Etch primers are not recommended, however, after phosphate pretreatments. As with ferrous metals, unless the anticipated exposure conditions are especially mild a further primer coat is applied once the etch primer has dried. On aluminium, this secondary primer is usually based on inhibitive zinc chromate. Where exposure conditions are moderate to severe, primers based entirely on this pigment, at a pigment volume concentration of about 35 per cent, are preferred; some extender material may also be added for specific reasons. In less demanding situations, red iron oxide–zinc chromate combinations can be used (page 107).

When stoving-cured paints are applied to aluminium in factory finishing there is no major restriction on the coatings that can be used, provided that stoving temperatures of about 120°C are not exceeded. This limiting temperature is variable, depending on the exact composition of the aluminium being coated, i.e. the base metal purity, alloying elements, etc.

7.3.3 PAINTS FOR ZINC

Zinc sheet, galvanised iron or sprayed zinc coatings have a moderate-to-high resistance to corrosion although painting is frequently required, not only for cosmetic reasons but also to reduce corrosion product formation on the surface of the zinc. This can reduce its service life as well as mar its appearance.

Etch primers are frequently used on galvanised or sprayed zinc coatings, followed by a zinc chromate primer, as for aluminium metal.

The traditional primers for zinc surfaces, and undoubtedly the best in respect of long-term durability with maintained adhesion, are those based on the inhibitive pigment calcium plumbate. Zinc tends to react with many oil-containing binders to form salts. Because these salts form at the substrate–film interface, they both reduce adhesion and cause embrittlement as a result of their acidic nature. Calcium plumbate primers appear to be more tolerant of these exacting conditions, and they are also suited for application to freshly galvanised steel where adhesion can be difficult to obtain. The toxicity of the pigment, however, does limit its use.

Topcoats can be based on conventional alkyd-based undercoat and gloss finish formulations as described earlier. In severe exposure conditions, however, the micaceous form of iron oxide is used in alkyd binder formulations. This particular pigment can be included in both the undercoat and topcoat formulations.

Micaceous iron oxide is a natural iron oxide with a crystalline flake-like structure. Especially when used in conjunction with small amounts of aluminium pigment, it produces low-permeability films of high durability with a reasonable degree of cosmetic appeal. The formulations of a micaceous oxide undercoat and topcoat are shown in Table 7.4, both having a pigment volume concentration of about 44 per cent.

Such formulations are also used on mild steel structures in severe exposure conditions, where this pigment can again be incorporated into all levels of the coating system. For steel, the undercoat formulation in Table 7.4 could be applied as a primer coat, followed by the topcoat.

Micaceous iron oxide paints are typically brush- or spray-applied and, for certain awkward situations such as painting electricity pylons, the paint must be capable of extreme dilution (to facilitate application) without loss of efficacy. These paint types are some of the few that are tolerant of such dilution.

Table 7.4 Micaceous Iron Oxide Paints

	Undercoat weight (%)	**Topcoat** weight (%)
Micaceous iron oxide	60·0	50·0
China clay	5·0	—
Leafing-grade aluminium paste, 70% powder in white spirit	1·5	15·0
Long-oil-length drying-oil-modified alkyd, 70% in white spirit	24·0	27·0
24% Lead naphthenate	0·35	0·4
6% Cobalt naphthenate	0·15	0·2
White spirit	8·4	6·8
Clay gellant	0·4	0·4
Industrial alcohol	0·2	0·2

7.4 Formulation of coatings for plastics

7.4.1 INTRODUCTION

Many types of plastics are available to modern industry. They can be produced self-coloured, but applied paint finishes are sometimes required for certain cosmetic or possibly protective reasons.

Plastics are bulk polymeric materials which therefore have certain similarities to paint coatings. They can be subdivided into two main types, thermoplastic and thermosetting resins. Plastic articles and components are basically fabricated by two main methods, moulding and forming.

Thermoplastic mouldings are typically produced by injection where the plastics material, made fluid by heating, is injected into a pretooled mould and allowed to set (cool). Thereafter, the mould is struck and the moulding is ejected. Thermosetting resin mouldings require a controlled heating and pressure regimen on a moulding press and, once set, these resultant plastics are irreversibly hardened. Thermoplastic components can also be made by forming, or fabrication, from sheet material.

7.4.2 PAINTS FOR PLASTICS

The nature of the plastics material, whether it is thermoplastic or thermosetting, has a marked influence on the solvent resistance of the resultant product, and naturally this influences the choice of formulation which can be applied. In general, *thermosetting* materials as a class are resistant to most common solvents; this group includes the melamine, urea, and phenol-formaldehyde types. Of the *thermoplastic* materials, nylon is resistant to most solvents, but other resins are attacked by specific classes of solvent. In particular, poly(vinyl chloride) (p.v.c.) is sensitive to ketones and esters as well as aromatic hydrocarbon solvents, as are acrylonitrile butadiene styrene (a.b.s.), polystyrene, and poly(methyl methacrylate) (acrylic) plastics. Alcohols and aliphatic hydrocarbons do not affect the above materials. Cellulose acetate plastics are attacked by ketones, esters and alcohols, but they are resistant to aromatic hydrocarbons. High-impact polystyrenes have similar resistance characteristics to polystyrene, but they are somewhat more tolerant of aromatic hydrocarbons.

Most paints can be applied to thermosetting plastics, and air-drying as well as stoving-cured formulations can be used. With the latter, however, the stoving temperature must obviously be limited to avoid the risk of distortion of the component. The use of stoving-cured paints on thermoplastic materials is limited by the very nature of the substrate itself.

The adhesion of paint to plastics also tends to vary, depending on the plastic itself and the binder component of the applied paint. A particularly difficult plastic to coat is polyethylene in both its low-density and high-density forms; polypropylene can also present problems.

With thermosetting plastics, surface roughening is required to remove the characteristic gloss finish obtained on striking the mould. Light sanding or wet-blasting can be used for this purpose. Because thermoplastic components are sensitive to solvents, surface roughening is not essential if a coating can be formulated using a solvent blend which slightly attacks (solubilises and etches) the surface of the substrate. This 'etching' promotes paint adhesion, although it generally necessitates spray application in order to ensure an even finish.

Plastics such as polystyrene, acrylic or cellulose acetate can be coated using alkyd or nitrocellulose paints with carefully selected solvent blends. Cellulose acetate plastics material is preferably coated with a paint based on cellulose acetate polymer, since nitrocellulose paints tend to soften as a result of plasticiser migration from the bulk resin.

The formulations of coatings for two widely-used industrial plastics materials are given below.

Poly(vinyl chloride)

Poly(vinyl chloride) in its unplasticised form is widely used for exterior situations in the building industry, especially for pipes, gutters, fittings, etc. Deterioration of this material during weathering results primarily in loss of colour; moreover, since articles are generally only available in greys or black, painting is frequently desirable for cosmetic reasons.

Because use is mainly in the building trade, application of coatings is typically by brush. Adequate performance can be obtained using an alkyd-based undercoat and topcoat without the need for a primer, although there is some advantage in allowing the p.v.c. to pre-weather before painting.

Table 7.5 Vinyl Finish for Poly(vinyl chloride) Plastics, White

	Weight (%)
Titanium dioxide, rutile	4·0
Vinyl chloride–vinyl acetate copolymer, ratio 87:13	16·0
Acetone	35·0
Cyclohexanone	10·0
Toluene	35·0

Two-pack etch primers can be used with such systems, however, and they may provide benefit in enhanced adhesion on unweathered surfaces.

Flexibilised p.v.c. mouldings are used for children's toys and finished industrially by spraying. Lacquers and paints based on vinyl copolymers are widely used on these plastics because of their high adhesion and flexibility. The pigment volume concentration of the formulation given in Table 7.5 is about 7 per cent, and would be adequate for this type of application involving children's toys.

The copolymers of poly(vinyl chloride) and poly(vinyl acetate), p.v.c.–p.v.a., are available in a range of ratios and, besides their use in coatings for plastics, they are widely used on exterior steelwork in severe exposure conditions. They are also used in applications such as the interior coating of tin-plate cans and containers. Copolymers of ratios 95:5 through to 87:13 are most commonly available for use in paints. Modification with about 1 per cent of a dibasic acid produces terpolymers with improved adhesion to metals.

Increasing the amount of vinyl acetate in the resin increases the range of solvents and diluents that can be used with the copolymer. Higher p.v.c. copolymers will accept little, if any, diluent and are most readily soluble in cyclic ketones. Lower p.v.c. copolymers are soluble in aliphatic ketones and will accept aromatic hydrocarbons as diluents. As a class, they tend to be poorly soluble in other solvent types.

With the formulation example shown in Table 7.5, speed of drying would be important in continuous production-line work; although the solvent blend used would enable rapid-drying sprayed coatings to be applied, some substitution of the cyclohexanone with a faster-evaporating ketone would further accelerate drying.

Polyesters

Polyester mouldings reinforced with glassfibre are used in the fabrication of such components as shells for boats and car bodies. Thorough surface preparation is again vital in order to obtain good adhesion of applied coatings. Typically, mould release agents have to be removed by a combination of detergent and white spirit washing.

A range of coatings can be used on polyester plastics and these are generally air-drying, or low-temperature force-drying. Primers are essential, however, in order to obtain good adhesion, and usually the slightly roughened surface of fibre-reinforced polyester necessitates that some filling action is also provided by the primer. This permits the surface to be sanded smooth prior to application of the topcoats. Typically, for car bodywork, a two-pack polyurethane primer–surfacer followed by one or more coats of an alkyd-nitrocellulose topcoat may be used. The general formulation principles of these types of finish have been discussed on page 120 and 141.

7.5 Bibliography

Newton, D.S. (Ed.), *Paint Technology Manual,* Part 4, *The Application of Surface Coatings.* Chapman and Hall: London, 1965.

Flick, E.W., *Water-based Paint Formulations.* Noyes Data Corporation: New Jersey, 1975.

Flick, E.W., *Solvent-based Paint Formulations.* Noyes Data Corporation: New Jersey, 1977.

Murphy, J.A. (Ed.), *Surface Preparation and Finishes for Metals.* McGraw-Hill: New York, 1971

Matthan, J. *et al.* (Eds.), *Ageing and Weathering of Plastics: A Review of the Literature.* Rubber and Plastics Research Association of Great Britain: Shrewsbury, 1970.

APPENDIX 1

Conversion Tables

Length Conversions

1 yard = 0.9144 m = 914.4 mm
1 foot = 0.3048 m = 304.8 mm
1 inch = 2.54 cm = 25.4 mm
1 milli-inch (thou) = 25.4 μm

1 metre = 1.0936 yd = 3.2808 ft
1 centimetre = 0.3937 in
1 millimetre = 0.039 in
1 micrometre = 0.039 thou

Foot (ft)	**Yard** (yd)	**Metre** (m)	**Metre** (m)	**Foot** (ft)	**Yard** (yd)
0·5		0·152	0·5	1·64	0·55
1		0·305	1	3·28	1·09
2		0·610	2	6·56	2·19
3	1	0·914	3	9·84	3·28
5		1·52	4	13·12	4·37
10		3·05	5	16·40	5·47
15		4·57	10	32·81	10·94
20		6·10	15	49·21	16·40
25		7·62	20	65·62	21·87
	2	1·83	25	82·02	27·34
	3	2·74	50	164·0	54·68
	4	3·66	100	328·1	109·4
	5	4·57	150	492·1	164·0
	10	9·14	200	656·2	218·72

Inch (in)	**Centimetre** (cm)	**Centimetre** (cm)	**Inch** (in)
0·5	1·27	1	0·39
1	2·54	2	0·79
2	5·08	3	1·18
3	7·62	4	1·57
4	10·16	5	1·97
5	12·70	10	3·94
6	15·24	15	5·91
8	20·32	20	7·87
10	25·40	25	9·84
12	30·48	50	19·69

Inch (in)	**Millimetre** (mm)	**Millimetre** (mm)	**Inch** (in)
0·25	6·35	1	0·039
0·5	12·70	2	0·079
0·75	19·05	5	0·197
1	25·4	10	0·39
2	50·8	15	0·59
3	76·2	20	0·79
4	101·6	25	0·98
5	127	50	1·97
10	254	100	3·94

Milli-inch (thou)	**Micrometre** (μm)	**Micrometre** (μm)	**Milli-inch** (thou)
0·1	2·54	1	0·04
0·25	6·35	2	0·08
0·5	12·7	3	0·12
0·75	19·1	4	0·16
1	25·4	5	0·20
2	50·8	10	0·39
3	76·2	15	0·59
4	101·6	20	0·78
5	127	25	0·98
10	254	50	1·95

Area Conversions

1 square yard = 0.8361 m^2
1 square foot = 0.0929 m^2 = 9.29 dm^2
1 square inch = 6.4516 cm^2

1 square metre = 1.196 yd^2 = 10.764 ft^2
1 square centimetre = 0.155 in^2

Square foot (ft^2)	Square metre (m^2)	Square metre (m^2)	Square foot (ft^2)	Square yard (yd^2)
0·5	0·046	0·5	5·38	0·60
1	0·093	1	10·76	1·20
2	0·19	2	21·53	2·39
3	0·28	3	32·29	3·59
4	0·37	4	43·06	4·78
5	0·46	5	53·82	5·98
10	0·93	10	107·6	11·96
15	1·39	15	161·5	17·94
20	1·86	20	215·3	23·9
25	2·32	25	269·1	29·9
50	4·65	50	538·2	59·8
75	6·97	75	807·3	89·7
100	9·29	100	1076	119·6
200	18·58	200	2153	238·2
300	27·87	300	3229	358·8
400	37·16	400	4306	478·4
500	46·45	500	5382	598
1000	92·90	1000	10,764	1196

Square inch (in^2)	Square centimetre (cm^2)	Square centimetre (cm^2)	Square inch (in^2)
1	6·45	10	1·55
5	32·26	25	3·88
10	64·52	50	7·75
20	129·03	100	15·50
30	193·55	150	23·25
40	258·06	200	31·00
50	322·58	250	38·75
100	645·16	300	46·50
150	967·74	400	62·00
200	1290·32	500	77·50

Volume Conversions

1 Imperial gallon = 4.5461 1
1 US gallon = 3.785 litres = 0.833 Imp. gal
1 imperial pint = 0.5683 1
1 fluid ounce = 28.413 cm^3

1 litre = 0.220 gal = 1.7596 pt
1 litre = 0.264 US gal
1 millilitre = 0.035 fl. oz = 1 cm^3

Pint (pt)*	**Gallon** (gal)*	**Litre** (l)	**Litre** (l)	**Pint** (pt)*	**Gallon** (gal)*
0·5		0·284	0·5	0·88	0·11
1		0·568	1	1·76	0·22
2		1·137	2	3·52	0·44
3		1·705	3	5·28	0·66
4	0·5	2·273	4	7·04	0·88
5		2·842	5	8·80	1·10
6		3·410	6	10·56	1·32
7		3·978	7		1·54
8	1	4·546	8		1·76
	2	9·092	9		1·98
	3	13·638	10		2·20
	4	18·184	15		3·30
	5	22·731	20		4·40
	10	45·461	50		11
	20	90·922	100		22
	50	227·305	250		55
	100	454·61	500		110

*Imperial

Fluid ounce (fl.oz)	**Millilitre** (ml)	**Millilitre** (ml)	**Fluid ounce** (fl.oz)
0·5	14·21	10	0·35
1	28·41	15	0·53
1·5	42·62	20	0·70
2	56·83	25	0·88
3	85·24	50	1·76
4	113·65	100	3·52
5	142·07	200	7·04
10	284·13	300	10·56
15	426·20	400	14·08
20	568·26	500	17·60

Imperial gallon	US gallon	US gallon	Imperial gallon
1	1·201	1	0·833
2	2·402	2	1·665
3	3·603	3	2·498
4	4·804	4	3·331
5	6·005	5	4·164
10	12·009	10	8·327
20	24·018	20	16·654
50	60·046	50	41·635
100	120·091	100	83·270

US gallon	Litre (l)	Litre (l)	US gallon
0·5	1·893	1	0·264
1	3·785	2	0·528
2	7·571	3	0·793
3	11·356	4	1·057
4	15·141	5	1·321
5	18·927	10	2·642
10	37·853	50	13·209
20	75·707	100	26·418
50	189·267	250	66·044
100	378·533	500	132·089

Temperature Conversions

$(°F-32)\times\frac{5}{9}=°C \qquad (°C\times\frac{9}{5})+32=°F$

°F →	°C	°C →	°F
0	–17·8	0	32
25	–3·9	5	41
32	0	10	50
50	10	15	59
75	23·9	20	68
100	37·8	25	77
125	51·7	50	122
150	65·6	75	167
175	79·4	100	212
200	93·3	125	257
212	100	150	302
225	107·2	175	347
250	121·1	200	392
300	148·9	250	482

Weight Conversions

1 pound = 0.4536 kg
1 ounce = 28.3495 g

1 kilogramme = 2.205 lb
1 gramme = 0.0353 oz

Pound (lb)	**Kilogramme** (kg)	**Kilogramme** (kg)	**Pound** (lb)
0·5	0·227	0·5	1·10
1	0·454	1	2·21
1·5	0·680	1·5	3·31
2	0·907	2	4·41
3	1·361	3	6·62
4	1·814	4	8·82
5	2·268	5	11·03
10	4·536	10	22·05

Ounce (oz)	**Gramme** (g)	**Gramme** (g)	**Ounce** (oz)
0·5	14·18	10	0·35
1	28·35	25	0·88
1·5	42·52	50	1·77
2	56·70	100	3·53
4	113·40	150	5·30
8	226·80	200	7·06
12	340·19	300	10·59
16	453·59	500	17·65

Concentrations

1 oz/gal* = 6.236 g/l   1 g/l = 0.1604 oz/gal

oz/gal	g/l	g/l	oz/gal
0·25	1·56	1	0·16
0·5	3·12	2	0·32
0·75	4·68	5	0·80
1	6·24	10	1·60
1·5	9·35	15	2·41
2	12·47	20	3·21
3	18·71	25	4·01
4	24·94	50	8·02
5	31·18	75	12·03
10	62·36	100	16·04
15	93·54	150	24·05
20	124·7	200	32·07
25	155·9	250	40·09
50	311·8	500	80·18
75	467·7	1000	160·36
100	623·6	2000	320·72

* Imperial gallon

APPENDIX 2

British Standards

There are numerous national and international standards that are relevant to paint technology. The following is a list of standards and codes of practice abstracted from the *British Standards Yearbook* issued by the British Standards Institution of London (1979 edition).

This list is not comprehensive but gives an indication of the range of composition and performance specifications available to the paint formulator.

Substrates and surface preparation

BS Code of Practice CP 231:1966 *Painting of buildings.*

BS Code of Practice CP 3012:1972 *Cleaning and preparation of metal surfaces.*

BS 4232:1967 *Surface finish of blast cleaned steel for painting.*

BS 5493:1977 *Code of practice for protective coating of iron and steel structures against corrosion.*

Coatings and allied products

BS 544:1969 *Linseed oil putty for use in wooden frames.*

BS 634:1961 *Air-drying insulating finishing varnish for electrical purposes.*

BS 1070:1973 *Black paint (tar-based).*

BS 1336:1971 *Knotting.*

BS 2521 and 2523:1966 *Lead based priming paints.*

BS 2524:1966 *Red oxide–linseed oil priming paints.*

BS 2778:1956 *Organic baking impregnating varnishes for electrical purposes.*

BS 3416:1975 *Black bitumen coating solutions for cold application.*

BS 3634:1963 *Black bitumen oil varnish.*

BS 3698:1964 *Calcium plumbate priming paints.*

BS 3761:1970 *Non-flammable solvent-based paint remover.*

BS 3826:1969 *Silicone-based water repellents for masonry.*

BS 4652:1971 *Metallic zinc-rich priming paint (organic media).*

BS 4756:1971 *Ready mixed aluminium priming paints for woodwork.*

BS 4764:1971 *Powder cement paints.*

BS 4842:1972 *Stoving organic finishes on aluminium extrusions and pre-formed sections for external architectural applications.*

BS 5082:1974 *Water thinned priming paints for wood.*

BS 5358:1976 *Specification for low-lead solvent-thinned priming paint for woodwork.*

FROM THE BRITISH STANDARDS AEROSPACE X SERIES:

3 X 17:1965 *Varnish for aeronautical purposes.*
2 X 18:1965 *Shellac varnish for aeronautical purposes.*
X 26:1966 *Doping and finishing schemes for fabric covered aircraft.*
X 27:1966 *Cellulose finishing scheme for aeronautical purposes.*
X 28:1966 *Matt synthetic finishing scheme for aeronautical purposes.*
X 29:1966 *High gloss finishing scheme for aeronautical purposes.*
X 30:1966 *Light weight aluminium finishing scheme (cellulose base) for aeronautical purposes.*
X 31:1967 *Low temperature stoving scheme for aeronautical purposes.*

Paint and raw material testing

BS 3483:1974 *Methods of testing pigments for paints.*
BS 3900:1965–1979 *Methods of test for paints.*
BS 3962:1965–1975 *Methods of test of clear finishes for wooden furniture.*
BS 4726:1971 *Methods of sampling raw materials for paints and varnishes.*

FROM THE BRITISH STANDARDS AUTOMOBILE SERIES

BS AU 148: 1969 *Methods of test for motor vehicle paints.*

Paint components: Binders

BS 311:1936 *Gold size for paints.*
BS 632:1969 *Linseed oil.*
BS 4725:1971 *Linseed stand oil.*

Paint components: Pigments and extenders

BS 217:1961 *Red lead for paints and jointing compounds.*
BS 239, 254, 296, 338, 637 and 1851:1967 *White pigments for paints.*
BS 282, 389:1963 *Lead chromes and zinc chromes for paints.*
BS 283:1965 *Prussian blues for paints.*
BS 284–6:1952 *Black (carbon) pigments for paints.*
BS 303:1965 *Brunswick or lead chrome greens (pure and reduced) for paints.*
BS 314:1968 *Ultramarine pigments.*
BS 318:1968 *Chromic oxide pigments.*
BS 388:1972 *Aluminium pigments.*
BS 1014:1975 *Pigments for Portland cement and Portland cement products.*
BS 1795:1976 *Specification for extenders for paints.*

BS 2876:1957 *Powder pigments for artists' use.*
BS 3599:1964 *Organic pigments for paints.*
BS 3699:1964 *Calcium plumbate for paints.*
BS 3981:1976 *Iron oxide pigments for paints.*
BS 3982:1966 *Zinc dust pigment.*
BS 4313:1968 *Strontium chromate for paints.*
BS 5193:1975 *Zinc phosphate pigments for paints.*

Paint components: Solvents and plasticisers

BS 135, 450, 805:1977 *Specifications for benzene, xylenes and toluenes.*
BS 245:1976 *Specifications for mineral solvents (white spirit and related hydrocarbon solvents) for paints and other purposes.*
BS 479:1963 *Aromatic naphthas.*
BS 508:1966 *n-Butanol.*
BS 509:1971 *Acetone.*
BS 551:1965 *n-Butyl acetate.*
BS 553:1965 *Ethyl acetate.*
BS 573:1973 *Plasticiser esters.*
BS 1940:1968 *Butanone (methyl ethyl ketone).*
BS 1941:1968 *4-methyl pentan 2 one (methyl iso butyl ketone).*
BS 2713:1966 *2-ethoxy ethanol.*

Miscellaneous

BS 332:1956 *Liquid driers for oil paints.*
BS 2015:1965 *Glossary of paint terms.*
BS 4310:1968 *Permissible limit of lead in low-lead paints and similar materials.*
BS 4800:1972 *Paint colours for building purposes.*

Index

Acetone 28, 173, 185
Acid catalyst 138, 139
Acid value 3
Acrylic acid 12
Acrylic resin 2, 7, 12, 26, 100, 102, 107, 131, 133, 135-136, 151, 167, 172
Adhesion 99, 104, 116, 127, 141, 149, 152, 164, 173
Alcohol 25, 28, 99, 118, 134, 154, 155, 171, 172
Aliphatic hydrocarbon 25, 106, 109, 116, 117
Alkali-resistance 146, 150, 152, 156, 160
Alkyd–amino finish 95-96, 101, 103, 138-140, 141
Alkyd resin 2, 4-6, 7, 26, 48, 55, 67, 95-97, 100, 108-109, 110, 111, 112, 113, 115, 129, 130, 133, 134, 135, 143, 153, 155, 159, 160, 168, 169, 171, 172, 174
Aluminium 28, 38, 40, 98, 169-170, 171, 183, 184
  paint 111-112, 131-132, 183
  painting of 122
  stearate 132, 147, 148
Amino resin 2, 6, 95-97, 104
Analysis of variance 85-87
Anatase, *see* Titanium dioxide
Anti-corrosion paint 112, 157
Anti-setting agent 32
Antimony oxide 16, 38, 40
Application, of paint 35, 45, 46, 69, 92, 106
  rate 69
Area conversion tables 177
Arithmetic mean 75-76
Aromatic hydrocarbons 25, 117, 118, 119, 122, 143, 172, 173, 185
Arylamide reds 20, 38, 40
Asbestine 113, 122. 123, 130
Average, *see* Mean
Azo dye 19, 20, 38, 40

Bactericide 31, 54, 131, 149, 150, 151, 152, 153, 165, 166
Baking, *see* Stoving cure
Ball mill 56, 105, 122
Barrier action 94
Barytes 23-24, 40, 101, 109, 117, 118, 120, 123, 131, 156, 160
Basic lead salts, *see* Lead
Bead mill 56
Bentonite 33
Benzidine yellow 19, 38, 40
Binder 2-14, 24, 25, 36, 53, 54, 120, 170
  solids 14, 35, 36, 67, 71, 136, 152
Binding ability 39, 157
Bingham body 47
Bioactive agent 30-32, 131
Bis-phenol A 6
Bitumen 13
Black pigments 18-19, 38, 40, 184

Blistering 37
Blue pigments 17-18, 19, 38, 40
Blush, *see* Solvent
Boiling point 27
British Standards 113, 130, 183-185
Brittleness 6, 8, 9, 28, 30
Brush application 26, 46, 108, 109, 112, 115, 116, 122, 131, 135, 148, 150, 157, 165, 170
Bubble formation 44
Butadiene 156, 157
Butanol 6, 27, 107, 139, 143, 185
Butanone, *see* Methyl ethyl ketone
Butyl acetate 28, 43, 142, 185
Butylated resin 6, 96, 138

Cadmium yellows 17, 38, 40
Calcium carbonate, *see* Chalk
Calcium plumbate 21, 38, 40, 114, 183, 185
Calcium stearate 147, 168
Calcium sulphate 145, *see* Gypsum
Carbon black 18-19, 38, 40, 97, 120, 133, 146, 167, 184
Carbon steel, *see* Steel
Castor oil 2, 3, 5, 48, 116, 117, 121, 123, 139, 155, 156, 157
Cellulose ether 33, 48
Cellulose resin 2, 10, 131, 140, 141-143, 172, 184
Cellulose thickener 150, 151, 152, 153, 165, 166, 167
Cement 145-160, 184
  paint 147, 183
Chalk 24, 40, 101, 109, 121, 130, 145, 156, 157
Chalking 61, 113
Chemical resistance 6, 7, 10, 103, 104, 105, 121
Chi-square test 87
China clay 24, 40, 101, 154, 165, 166, 171
Chlorinated paraffin 29, 116, 117, 118, 154, 155, 156, 157
Chlorinated rubber 2, 10, 26, 29, 44, 65, 93, 103, 112, 115, 116, 117, 118, 120, 145, 153, 155, 156, 157, 159, 165
Chrome greens 17, 38, 40, 61, 184
Chromium oxide 17, 38, 40, 122, 146, 184
Clays 33, 40, 48, 101, 107, 108, 116, 118, 120, 134, 145, 154, 155, 157, 168, 171
Clear finish 104, 137, 139, 140, 142, 143, 184
Coal tar epoxy 7, 13
Coalescing solvent 28, 57, 131, 135, 136, 149, 151, 152, 153, 166
Colbalt drier 3, 30, 67, 97, 107, 109, 110, 112, 113, 130, 132, 134, 137, 140, 141, 160, 168, 171
Coefficient of variation 77
Cold cure 7
Colour 14, 53, 58-61, 113
Colouring pigment 15-20, 38, 40, 184
Concentration conversions 181
Concrete 145-160
Conversion tables 175-181
Convertible coating 2-10, 25
Copal resin 3

Copolymer 5, 11, 12
Correlation 87-88
Corrosion-inhibitive pigment 20-22, 38, 40, 94, 114, 115
Corrosion protection 14, 22, 93-94
Corrosion resistance 37
Coumarone 111, 112, 132
Critical pigment volume concentrations (C.P.V.C.) 37-42, 72
Cross-linkage 7, 9
Curing agent 119, 121
Cyclohexanol 103

Daniel flow point 56, 57
D.C.O., *see* Castor oil
Defoamer 131
Degrees of freedom 83
Density, *see* Specific gravity
Dibutyl phthalate 29-30, 142, 143, 157
Dichlorofluamide 31
Diluent 44, 106, 119, 122, 142, 143, 173
Di-octyl phthalate 30
Dispersing agent 33, 56, 148
Dispersion, *see* Pigment
Dispersion coating 104-105, 106, 148
Dispersion of data 77
Dithiocarbamate 31
Driers 30, 66-67, 110, 134, 185
Drying oil 2-4, 5, 13, 108, 112
Drying time 3, 26, 30, 35, 42, 43, 104, 118, 130
Durability 3, 5, 8, 37, 53, 94, 101, 103, 110, 115, 120, 130, 141, 142, 151, 153, 157

Efflorescence 164-165
Egg-shell finish 25, 58, 167
Electrophoretic painting 92
Electrostatic painting 92
Emulsion paint 24, 30, 37, 41, 56, 57, 58, 64, 129, 131, 133, 134-136, 137, 146, 148, 149, 151, 153, 159, 164, 165, 166, 167
Epoxy ester 7, 97, 100
Epoxy resin 2, 6-7, 25, 26, 44, 99, 103, 104, 115, 119-120, 121, 122, 123, 159
Epoxy–phenolic 100, 103-104, 106
Ester 25, 28, 119, 143, 157, 172, 185
Etch primer 12, 98, 99-100, 169
Ethanol, *see* Alcohol
Ether 25, 28, 43, 119, 123, 153
Ethoxyacetate 118, 121, 155, 156
Ethyl acetate 28, 122, 185
Evaporation rate 25, 26, 27, 43-44, 60
Expectation (in probability theory) 79-80
Exposure conditions 93
Extender 1, 14-15, 23-24, 33, 36, 40, 101, 122, 132, 150, 157, 160
Exterior cement/concrete 147-160
Exterior finish 8, 37, 39, 150, 151, 153
Exterior timber 129
Exterior wood stain 132-136

F-test 85
Fatty acid 4, 7, 54
process 4
Filler 128
Film-build 94, 115, 117, 118, 133, 134, 140, 143, 151, 154, 155, 157
Film flexibility 3, 12, 120, 122, 130, 131
Film thickness 69, 135
Finish 96, 97, 98, 101, 102, 104, 108, 109, 112, 116, 117, 120, 121, 123, 135, 159-160, 170, 174, 184
Flash-point 27, 45, 99
Flatting 24, 142
Flooding 60
Flow 3, 32-33, 43, 46, 48, 49, 122, 123
Formulation, of paint 35-61
Fouling 31
Fungal growth 31, 134, 149
Fungicide 31, 134, 135, 149

Galvanised iron 168, 170
Gellant, *see* Thickener
Glass transition temperature 7, 29
Gloss 6, 23, 41, 53, 58, 94, 95, 101, 108, 109, 122, 142, 143, 172, 184
Glycerol 4, 5, 8
Green pigments 17, 19, 38, 40
Gypsum 147, 163-168

Hammer finish 97, 98
Hansa yellows 19, 38, 142
Hardness 7
Hardwood 125, 126, 128, 131
Heartwood 126
Hegman gauge 58
Helio reds 20, 38, 142
High film-build, *see* Film-build
Hydrogen bonding 25
Hydrogenated castor oil, *see* Castor oil

Immersion coating 92
Industrial coatings, for steel 94-123
Inhibitor, *see* Corrosion
Inorganic silicate 13, 115, 120
Interior paint 37, 165, 168
Interior timber 136-143
Iron 91
Iron carbide 91
Iron oxide
black 18, 38, 40
brown 133
micaceous 115, 146, 170, 171
red 18, 38, 40, 100, 107, 108, 114, 122, 123, 130, 132, 133, 134, 135, 169, 185
scale 92
yellow 17, 38, 40, 133
Isocyanate 6, 8-9, 112, 120, 121, 123

Kaolin 24, 40
Kauri resin 3
Ketones 25, 119, 122, 143, 172, 173
Kieselghur 130
Knots 132, 183

Lacquer 138-143, 173
Lead 23, 38, 40, 67, 106, 183, 185
  chromate 16, 38, 40, 156, 184
  chrome green 17, 38, 40, 61, 184
  drier 3, 107, 109, 110, 111, 116, 117, 130, 132, 134, 137, 160, 168, 171
  primer 113-114, 116, 117, 129-130
  silicochromate 18, 20, 38, 40, 116, 117
  sulphate, basic 21, 38, 40
Length conversion tables 175-176
Lightfastness 60-61
Lignin 126
Lime 145, 147
Linoleate 30
Linseed oil 2-4, 42, 108, 113-114, 129, 130, 183, 184

Magnetite 18, 38, 40
Maintenance paint 112, 115
Manganese drier 30, 97
Mann-Whitney test 88
Marine paint 7, 118, 119, 137
Masonry paint 37, 145-160
Matt finish 150, 153, 167
Mean 75-76
  deviation 76
Median 75
Melamine resin 6, 95, 101, 104, 138, 171
Metallic effect 97, 111-112
Metallic pigments 22-23, 38, 40, 111, 115
Metamerism 59
Methacrylic acid 12
Methyl ethyl ketone 28, 121, 123, 143, 185
Methyl iso-butyl ketone 28
Mica 24, 60, 109, 115, 117, 120, 131, 138, 152, 155, 156, 157
Micaceous iron oxide 115, 146, 170, 171
Mild exposure, steel 94-97, 107-112
Mill base 56, 57, 58
Mode 75
Moderate exposure, steel 99-103, 112-114
Modifying agent 99
Moisture cure 9

Naphthanate, *see* Lead and Cobalt driers
Newtonian fluid 46, 49
Nitrocellulose 10, 27, 101, 141-143, 172, 174, 184
Non-convertible binder 2, 10-14, 25, 26
Non-drying oil 2, 6, 142
Non-ferrous metals 168-171
Non-parametric tests 88

Normal curve 75, 78-80
Nylon 171

Octoates 30
Oil absorption 40, 41-42, 56, 73, 106, 142
Oil length 4, 8, 130, 137, 168
Oils 2-4
Oleoresinous varnish 2-4, 7, 130, 132, 137
Olive oil 2
Opacity 14, 114, 151, 158, 160, 166, 167
Organomercurials 32, 134
Organosol 11, 103, 104-105
Organotin compounds 32, 106, 127

Paint formulae 63-65
Paint formulation 35-61, 64
  non-statistical method 51-53
  statistical method 51
Paint manufacture 56-58, 96
Paint requirement 70
Penetration of paint 127-129
Percentage volume solids 67-68, 71
Percentage weight 36, 64
Permeability 29, 37, 39, 41, 111, 114, 120
Phenolic resin 2, 7, 8, 100, 132
Phloem 125
Phthalic acid 4
Phthalic anhydride 108
Phthalocyanine blue 19, 38, 40, 120, 146
Pigment 1, 14-23, 32, 33, 40, 72, 94, 101, 106, 118, 119, 133, 148, 157, 184
  addition 38
  –binder ratio 36-37, 56, 70-71, 99
  dispersion 3, 53-56, 58, 59
  green B 19, 38, 40
  loading 36-42, 166
  volume concentration 35, 36, 37-42, 71-73, 107, 109, 110, 114, 115, 116, 118, 123, 130, 132, 133, 135, 138, 139, 141, 149, 150, 151, 153, 155, 156, 157, 160, 165, 166, 167, 169, 170
Pink primer 129, 130, 183
Plaster, *see* Gypsum
Plastic, painting of 171-174
Plasticiser 6, 28-30, 104-105, 116, 118, 141, 157, 172
Plastisol 11, 103, 104-105
P.M.M.A., *see* Acrylic resin
Poise 49, 50
Polyester 4, 5, 26, 138, 140, 141, 173-174
Polyethylene 172
Polyhydric alcohol 4, 122
Polymer, *see* Resin
Polypropylene 172
Polystyrene, *see* Styrene
Polyurethane resin 2, 8-9, 10, 26, 44, 115, 119, 120-123, 128, 137, 153, 159, 174
Poly(vinyl butyral) 12, 98, 99, 103
Porosity 37

Portland cement 145, 184
Precipitation point 44
Pretreatment primer, *see* Etch primer
Primer 54, 58, 109, 183
  for polyesters 174
  for steel 2, 37, 94, 97, 100, 107-108, 113-114, 120-123
  for timber 2, 129-132
Probability 78-80
  level 80
Prussian blue 17, 38, 40, 60, 122, 156, 184
p.v.c., p.v.a., *see* Vinyl resins
P.V.C., *see* Pigment volume concentration

Range 75
Rays, in timber 125, 126, 128
Red iron oxide 18, 38, 40, 100, 107, 108, 114, 122, 123, 130, 132, 133, 134, 135, 169
Red lead 18, 21, 38, 40, 113-114, 122, 129, 130, 183, 184
Red pigments 18, 19-20, 38, 40
Refined oil 3
Reflow coating 102
Regression analysis 87-88
Resin 2, 25, 65, 66, 141, 143
Rivelling 30, 116
Roller application 46, 165
Rosin ester gum 4
Rosin modification 8
Rutile, *see* Titanium dioxide

Sand 145, 151
Sand mill 56
Sapwood 126
Scrub resistance 41, 94
Sealer 128, 142, 158-159, 164, 165
Seasoning of timber 126-127
Self-etch primer, *see* Etch primer
Semi-drying oil 2-4, 95, 108, 168
Semi-gloss finish 41, 58, 135, 153, 160, 167
Settling, of pigments 32, 111
Severe exposure, of steel 103-106, 115-123
Significance, statistical 80-81
Silica 33, 55, 95, 97, 122, 140, 157, 168
Silicate 13
Silicone fluid 97, 98, 123
Silicone resin 2, 9-10, 112, 113, 183
Siloxane 9
Site-applied coatings 107-123
Soap 3, 149
Sodium carboxy methyl cellulose 33, 48
Softwood 125, 129, 132
Solubility parameter 25, 26
Solvent 1, 11, 24-28, 42-45, 56, 96, 106, 117, 118, 119, 122, 142, 171, 173, 185
  blush 43
  power 44, 143
  resistance 25, 171

Soya bean oil 2, 5, 95, 108, 137, 139
Specific gravity
  of paints 64, 65-67
  of pigments 15, 23, 40, 42, 60
  of polymers 14
  of solvents 27
Spraying of paint 44, 46, 92, 97, 112, 118, 122, 140, 155, 156, 170, 172, 173
Stabiliser 106
Stain, *see* Wood stain
Stain resistance 41, 94
Stainless steel 22, 38, 40, 115
Standard deviation 76, 83
Standard error of the mean 81-83
Standard score 77, 78, 80-81, 82
Statistical tests 78, 80-88
Statistics 73-88
Steel 91
  painting of 92-123
Stoke 49, 50
Stoving cure 5, 6, 7, 8, 11, 97, 100, 102, 104, 170, 172, 183
Student's *t* test 83-85, 87, 88
Styrenated resin 5, 6, 112, 151
Styrene 5, 107, 140, 141, 156, 172
Surface-active agent 33-34, 54-56, 148, 152, 165, 166
Swimming pool coating 155

t-statistic, *see* Student's *t* test
Talc 24, 40, 101, 117, 120, 123, 150, 151
Temperature conversion table 179
Textured paint 151-152, 159
Thermal stability 7
Thermoplastic resin 11, 12, 171, 172
Thermosetting resin 12, 102, 171, 172
Thickener 32, 45, 48, 56, 101, 116, 134, 135, 136, 148, 158
Thixotropic paint 32, 46-47, 140, 168
Timber, painting of 125-143
Tint base 150, 166-167
Tint strength 59-60, 148
Titanium dioxide 15-16, 38, 40, 55, 61, 96, 101, 102, 105, 109, 117, 118, 120, 123, 131, 139, 146, 147, 150, 151, 153, 154, 155, 156, 157, 159, 160, 165, 168, 173
Toluene 27, 123, 173, 185
Toluidine reds 19-20, 38, 40
Toner 19, 156
Top coat, *see* Finish
Tracheids 125, 128
Tri-butyl tin oxide 127
Tung oil 2, 3, 96, 132, 137, 159

U-statistic 88
Ultramarine blue 17-18, 38, 40, 156, 184
Ultra-violet light 128, 131, 133, 137, 138
Undercoat 108, 109, 112, 115, 116, 135, 169, 170

Urea-formaldehyde resin 6, 95, 104, 138, 139, 171
Urethane–alkyd 6
Urethane resin, *see* Polyurethane

Variance 76, 84
analysis of 85-87
ratio 85
Variation 76-78
coefficient of 77
Varnish 2-4, 111, 130, 132, 133, 136-138, 153, 157, 159, 184
Vegetable oil 2, 5, 7, 108, 153, 154, 157
Vinyl resin 2, 6, 10-12, 26, 29, 44, 93, 97, 100, 103, 106, 112, 115, 140, 146, 152, 153, 156-157, 159, 165, 166, 167, 172, 173
Vinylidene resin 12
Viscosity 3, 24, 32-33, 45-49, 96, 155
control 81, 82, 83, 101
measurement 49, 50
Volume additon rate 38, 67-68
Volume conversion tables 178-179

Wash primer, *see* Etch primer
Water resistance 99, 121, 153
Water vapour permeability 41
Weathering
of cement and concrete 146, 158
of timber 126
Weight conversion tables 180
Wetting agent, *see* Surface-active agent
White lead 16, 38, 40, 129, 130
White pigments 15-16, 38, 40, 60, 61, 184
White spirit 27, 107, 109, 112, 113, 115, 117, 132, 134, 137, 154, 155, 157, 158, 159, 160, 168, 171, 174, 185
Whiting 24, 40, 131, 151, 165, 166, 168
Wilcoxon test 88
Wood, *see* Timber
Wood stain 132-136
Wrinkle finish 96, 97

Xylem 125, 126
Xylene 27, 96, 112, 118, 121, 139, 155, 156, 159

Yellow pigments 16-17, 19, 38, 40, 61
Yellowing 168

Z-score, *see* Standard score
Zinc 13, 23, 38, 40, 116, 120, 121, 168, 169, 170, 183, 185
chromates 16-17, 38, 40, 98, 100, 107, 108, 114, 122, 123, 169, 184
oxide 16, 32, 38, 40, 122
phosphate 20, 21, 38, 40, 114, 185
potassium chromate 21
tetroxychromate 21-22, 38, 40, 98, 99
Zirconium drier 30, 110